SRA

SPECTRUM MATH

red

FOURTH EDITION

AUTHOR
Thomas J. Richards
Mathematics Teacher
Lamar Middle School
Lamar, Missouri

SRA/McGraw-Hill
Columbus, Ohio

Photo Credits
Ronnie Kaufman/The Stock Market, **2;** Doug Martin/SRA/McGraw-Hill, **50, 80;** Robert Brenner/PhotoEdit, **60;** David Young-Wolff/PhotoEdit, **88;** Jeff Zaruba/Tony Stone Images, **102;** Tony Freeman/PhotoEdit, **124.**

SRA/McGraw-Hill

A Division of The **McGraw-Hill** *Companies*

Send all inquiries to:
SRA/McGraw-Hill
8787 Orion Place
Columbus, OH 43240-4027

Printed in the United States of America.

ISBN 0-02-687543-8

6 7 8 9 POH 04 03 02 01

Contents

Chapter 8

Multiplication
(basic facts through 9 × 9)

Chapter 9

Multiplication
(2-digit by 1-digit)

Chapter 10

Division
(basic facts through 45 ÷ 5)

Chapter 11

Division
(basic facts through 81 ÷ 9)

Chapter 12

Metric Measurement

Chapter 13

Customary Measurement

NAME ____________________

Readiness Check

Add.

$$\begin{array}{r} 4 \\ +5 \\ \hline \end{array} \quad \begin{array}{r} 8 \\ +7 \\ \hline \end{array} \quad \begin{array}{r} 20 \\ +40 \\ \hline \end{array} \quad \begin{array}{r} 31 \\ +52 \\ \hline \end{array} \quad \begin{array}{r} 86 \\ +3 \\ \hline \end{array} \quad \begin{array}{r} 76 \\ +17 \\ \hline \end{array}$$

$$\begin{array}{r} 7 \\ 6 \\ +5 \\ \hline \end{array} \quad \begin{array}{r} 20 \\ 40 \\ +30 \\ \hline \end{array} \quad \begin{array}{r} 13 \\ 25 \\ +41 \\ \hline \end{array} \quad \begin{array}{r} 400 \\ +300 \\ \hline \end{array} \quad \begin{array}{r} 642 \\ +345 \\ \hline \end{array} \quad \begin{array}{r} 702 \\ +146 \\ \hline \end{array}$$

Subtract.

$$\begin{array}{r} 9 \\ -3 \\ \hline \end{array} \quad \begin{array}{r} 10 \\ -6 \\ \hline \end{array} \quad \begin{array}{r} 15 \\ -7 \\ \hline \end{array} \quad \begin{array}{r} 40 \\ -30 \\ \hline \end{array} \quad \begin{array}{r} 68 \\ -23 \\ \hline \end{array} \quad \begin{array}{r} 56 \\ -26 \\ \hline \end{array}$$

$$\begin{array}{r} 67 \\ -30 \\ \hline \end{array} \quad \begin{array}{r} 91 \\ -16 \\ \hline \end{array} \quad \begin{array}{r} 53 \\ -47 \\ \hline \end{array} \quad \begin{array}{r} 60 \\ -23 \\ \hline \end{array} \quad \begin{array}{r} 184 \\ -62 \\ \hline \end{array} \quad \begin{array}{r} 567 \\ -140 \\ \hline \end{array}$$

Write the next three numbers.

23, 24, 25, ________ , ________ , ________

87, 88, 89, ________ , ________ , ________

137, 138, 139, ________ , ________ , ________

Count by 10.

30, 40, 50, ________ , ________ , ________ , ________ , ________

400, 410, 420, ________ , ________ , ________ , ________ , ________

170, 180, 190, ________ , ________ , ________ , ________ , ________

Readiness Check (continued)

Ring the fraction that tells how much is blue.

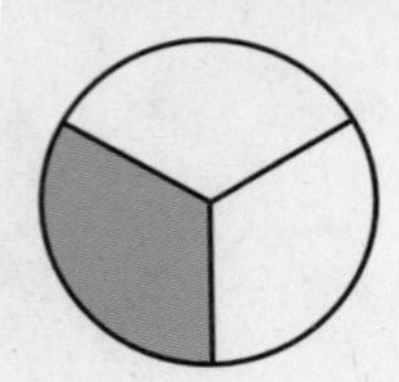

$\frac{1}{2}$ $\frac{1}{3}$ $\frac{1}{4}$

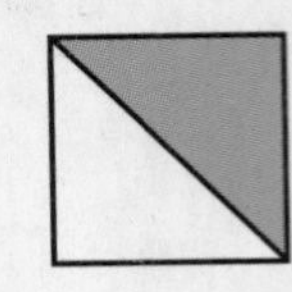

$\frac{1}{2}$ $\frac{1}{3}$ $\frac{1}{4}$

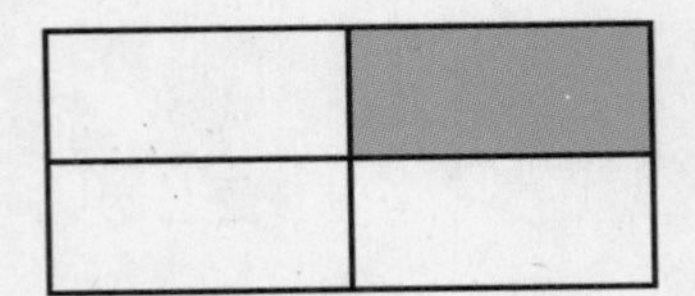

$\frac{1}{2}$ $\frac{1}{3}$ $\frac{1}{4}$

Write the time for each clock.

_____ : _____

_____ : _____

_____ : _____

How long is each object?

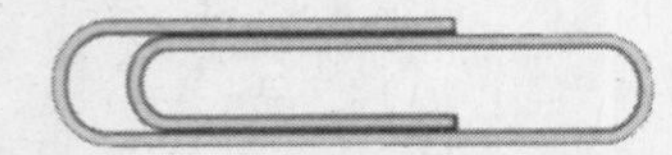

_____ centimeters

_____ inches

Solve each problem.

Mick has 346 pennies.

Matt has 213 pennies.

How many pennies do they have in all?

Ms. Smith had 367 dollars.

She spent 132 dollars.

How many dollars does she have left?

NAME ____________________

Addition Facts (Form A)

	a	b	c	d	e	f	g	h
1.	3 +1	8 +2	1 +6	4 +7	6 +3	2 +8	4 +5	7 +9
2.	6 +4	1 +8	3 +9	2 +1	5 +0	0 +2	9 +1	3 +2
3.	2 +7	6 +9	4 +8	9 +3	2 +2	8 +0	0 +4	7 +1
4.	5 +2	8 +3	1 +5	7 +8	6 +2	4 +6	5 +4	9 +4
5.	2 +3	9 +0	4 +3	2 +9	1 +1	8 +8	3 +5	5 +7
6.	8 +9	3 +3	9 +5	6 +6	3 +8	0 +6	7 +3	2 +6
7.	7 +7	4 +1	3 +6	8 +7	0 +0	9 +8	9 +2	7 +5
8.	2 +4	0 +3	5 +8	2 +5	1 +9	1 +0	5 +9	8 +4
9.	6 +7	3 +4	9 +9	0 +7	8 +5	7 +4	5 +6	3 +7
10.	9 +7	8 +6	5 +5	7 +6	6 +8	6 +5	4 +9	9 +6

NAME ____________________

Addition Facts (Form B)

	a	*b*	*c*	*d*	*e*	*f*	*g*	*h*
1.	8 +2	7 +0	0 +1	1 +1	6 +4	5 +2	4 +9	2 +7
2.	1 +0	6 +3	3 +0	2 +3	7 +1	8 +1	6 +5	1 +9
3.	0 +5	1 +2	6 +6	3 +5	9 +5	5 +7	7 +6	3 +8
4.	4 +2	6 +8	8 +5	2 +6	5 +8	9 +8	0 +0	4 +4
5.	7 +2	9 +7	0 +8	4 +7	7 +9	5 +9	3 +3	5 +4
6.	1 +3	9 +0	2 +2	5 +1	7 +7	6 +0	8 +6	9 +4
7.	4 +8	9 +3	1 +4	2 +9	9 +2	8 +3	7 +3	0 +9
8.	2 +0	2 +8	8 +4	4 +0	8 +7	9 +1	4 +3	5 +5
9.	8 +9	5 +6	6 +1	1 +7	4 +6	7 +5	9 +9	6 +7
10.	3 +9	9 +6	7 +8	5 +3	6 +9	8 +8	7 +4	3 +7

NAME ______________________

Subtraction Facts (Form A)

	a	*b*	*c*	*d*	*e*	*f*	*g*	*h*
1.	11 −3	8 −4	5 −5	12 −3	2 −1	10 −9	4 −3	11 −9
2.	10 −5	3 −3	6 −3	11 −4	7 −6	10 −6	9 −2	12 −4
3.	16 −7	9 −0	5 −4	13 −7	10 −2	15 −9	8 −8	14 −5
4.	13 −8	4 −2	7 −7	12 −9	2 −0	17 −9	6 −1	11 −7
5.	18 −9	9 −8	6 −4	11 −5	3 −1	15 −7	9 −9	10 −8
6.	12 −6	8 −7	3 −2	13 −9	10 −4	14 −6	7 −5	12 −5
7.	15 −8	8 −3	9 −5	12 −8	8 −6	16 −9	5 −3	12 −7
8.	14 −7	7 −1	6 −5	11 −6	4 −1	10 −7	1 −1	10 −3
9.	13 −4	0 −0	8 −0	16 −8	9 −7	14 −9	6 −6	13 −6
10.	17 −8	9 −6	7 −4	15 −6	11 −2	13 −5	9 −3	14 −8

NAME ______________________________

Subtraction Facts (Form B)

	a	b	c	d	e	f	g	h
1.	4 − 2	13 − 7	3 − 2	10 − 1	6 − 5	8 − 1	14 − 5	10 − 7
2.	8 − 2	12 − 5	6 − 3	10 − 8	2 − 1	11 − 9	14 − 8	11 − 2
3.	4 − 0	11 − 3	9 − 1	15 − 6	5 − 0	7 − 1	13 − 8	10 − 9
4.	6 − 4	13 − 9	1 − 0	9 − 2	7 − 3	12 − 4	15 − 7	5 − 4
5.	0 − 0	12 − 3	8 − 4	14 − 6	8 − 5	10 − 4	16 − 9	11 − 6
6.	9 − 9	10 − 2	3 − 2	15 − 9	5 − 1	12 − 9	14 − 9	10 − 3
7.	7 − 5	12 − 7	7 − 0	14 − 7	7 − 2	11 − 4	16 − 7	11 − 5
8.	4 − 4	13 − 6	5 − 2	16 − 8	9 − 4	10 − 5	13 − 4	6 − 0
9.	8 − 3	12 − 6	1 − 1	18 − 9	4 − 3	12 − 8	14 − 6	13 − 5
10.	9 − 6	11 − 7	8 − 8	17 − 8	6 − 2	10 − 6	17 − 9	15 − 8

NAME ______________________________

Multiplication Facts (Form A)

	a	*b*	*c*	*d*	*e*	*f*	*g*	*h*
1.	2×2	6×3	0×1	3×2	8×0	1×1	7×1	8×4
2.	7×4	3×0	8×3	2×1	5×1	3×6	2×5	6×2
3.	3×7	5×5	8×6	6×0	4×9	9×1	7×2	4×3
4.	8×5	2×4	7×5	4×1	8×2	6×5	7×8	1×9
5.	4×0	8×1	9×3	5×6	3×8	2×9	5×7	9×2
6.	7×9	6×4	4×8	7×3	6×9	9×4	2×6	8×7
7.	1×3	9×5	5×3	8×8	4×5	0×7	3×4	7×6
8.	3×5	9×0	2×7	7×7	5×8	9×6	2×0	6×6
9.	9×9	1×8	6×8	0×0	9×7	0×5	3×9	8×9
10.	4×6	9×8	2×8	4×7	1×6	6×7	3×3	5×9

NAME ______________________________

Multiplication Facts (Form B)

	a	*b*	*c*	*d*	*e*	*f*	*g*	*h*
1.	2 ×2	6 ×4	4 ×1	7 ×7	1 ×0	2 ×9	0 ×2	1 ×4
2.	5 ×0	3 ×1	8 ×3	3 ×9	0 ×9	9 ×6	7 ×3	4 ×8
3.	1 ×2	4 ×9	6 ×3	7 ×2	5 ×7	1 ×5	2 ×8	8 ×2
4.	8 ×4	9 ×5	4 ×2	2 ×3	6 ×9	4 ×7	4 ×3	5 ×6
5.	2 ×4	7 ×1	3 ×3	6 ×2	9 ×4	5 ×1	6 ×8	7 ×6
6.	7 ×4	1 ×1	5 ×8	8 ×5	7 ×0	0 ×8	6 ×1	9 ×3
7.	3 ×8	9 ×7	5 ×2	2 ×6	3 ×4	8 ×9	7 ×5	6 ×7
8.	8 ×6	5 ×4	4 ×6	9 ×2	1 ×7	8 ×1	4 ×4	3 ×5
9.	6 ×5	7 ×8	9 ×9	0 ×0	3 ×7	5 ×5	2 ×5	8 ×8
10.	9 ×1	4 ×5	8 ×7	6 ×6	9 ×8	7 ×9	2 ×7	5 ×9

NAME ____________________

Division Facts (Form A)

	a	b	c	d	e	f	g
1.	7) 7	4) 24	9) 18	3) 18	8) 32	6) 12	2) 8
2.	8) 0	1) 9	5) 15	2) 16	7) 21	5) 0	8) 8
3.	3) 15	8) 40	7) 28	4) 20	7) 63	3) 21	9) 36
4.	7) 14	5) 20	6) 6	2) 18	6) 24	1) 2	2) 10
5.	8) 24	5) 10	4) 28	9) 45	1) 8	5) 45	8) 48
6.	5) 40	6) 30	1) 6	5) 5	9) 0	8) 16	4) 4
7.	9) 54	1) 5	7) 56	6) 18	4) 16	6) 54	3) 6
8.	7) 35	3) 12	2) 0	8) 56	2) 12	6) 0	7) 49
9.	4) 0	8) 64	5) 35	4) 32	3) 24	1) 3	6) 36
10.	6) 42	9) 9	4) 8	1) 0	9) 63	4) 12	5) 25
11.	3) 9	2) 14	9) 72	7) 42	2) 4	8) 72	1) 1
12.	9) 81	6) 48	4) 36	2) 6	5) 30	1) 4	3) 27

NAME ____________________

Division Facts (Form B)

	a	b	c	d	e	f	g
1.	$2\overline{)2}$	$4\overline{)12}$	$3\overline{)9}$	$6\overline{)24}$	$8\overline{)48}$	$3\overline{)6}$	$8\overline{)0}$
2.	$6\overline{)30}$	$9\overline{)36}$	$7\overline{)14}$	$2\overline{)4}$	$5\overline{)5}$	$5\overline{)40}$	$7\overline{)63}$
3.	$1\overline{)7}$	$5\overline{)0}$	$5\overline{)45}$	$9\overline{)45}$	$4\overline{)8}$	$1\overline{)9}$	$8\overline{)56}$
4.	$3\overline{)3}$	$4\overline{)16}$	$7\overline{)56}$	$5\overline{)35}$	$8\overline{)8}$	$4\overline{)4}$	$9\overline{)54}$
5.	$7\overline{)0}$	$3\overline{)12}$	$8\overline{)64}$	$6\overline{)36}$	$7\overline{)21}$	$2\overline{)6}$	$4\overline{)36}$
6.	$9\overline{)27}$	$2\overline{)8}$	$6\overline{)18}$	$9\overline{)0}$	$6\overline{)54}$	$1\overline{)0}$	$6\overline{)12}$
7.	$6\overline{)0}$	$4\overline{)20}$	$8\overline{)40}$	$1\overline{)1}$	$8\overline{)72}$	$3\overline{)15}$	$5\overline{)30}$
8.	$9\overline{)18}$	$5\overline{)25}$	$7\overline{)49}$	$4\overline{)24}$	$3\overline{)24}$	$9\overline{)63}$	$2\overline{)10}$
9.	$3\overline{)0}$	$9\overline{)9}$	$6\overline{)48}$	$2\overline{)14}$	$6\overline{)6}$	$1\overline{)6}$	$8\overline{)16}$
10.	$3\overline{)18}$	$7\overline{)35}$	$1\overline{)4}$	$9\overline{)72}$	$4\overline{)28}$	$2\overline{)12}$	$7\overline{)42}$
11.	$1\overline{)8}$	$8\overline{)32}$	$5\overline{)20}$	$5\overline{)10}$	$2\overline{)18}$	$6\overline{)42}$	$5\overline{)15}$
12.	$8\overline{)24}$	$3\overline{)21}$	$9\overline{)81}$	$2\overline{)16}$	$7\overline{)28}$	$3\overline{)27}$	$4\overline{)32}$

NAME ____________________

Mixed Facts

Add, subtract, multiply, or divide. Watch the signs.

	a	*b*	*c*	*d*	*e*	*f*	*g*
1.	$7 + 6$	$8 + 3$	$14 - 9$	9×9	$13 - 6$	7×9	$5\overline{)45}$
2.	3×4	$9 + 7$	$11 - 7$	6×4	$9 + 9$	$10 - 6$	$6\overline{)42}$
3.	$15 - 8$	5×5	4×7	$12 - 5$	$7 + 8$	$5 + 4$	$8\overline{)64}$
4.	$7 + 5$	$16 - 8$	$4 + 8$	$14 - 7$	9×0	6×6	$7\overline{)35}$
5.	7×7	$9 + 5$	$17 - 8$	5×6	$8 + 9$	$17 - 9$	$9\overline{)63}$
6.	$15 - 6$	$13 - 9$	$5 + 6$	$8 + 5$	9×6	4×8	$7\overline{)56}$
7.	$6 + 6$	3×9	8×8	$7 + 7$	$18 - 9$	$10 - 5$	$8\overline{)40}$
8.	7×8	$16 - 9$	$11 - 3$	$2 + 9$	$8 + 8$	6×3	$9\overline{)81}$

NAME ______________________

Mixed Facts

Add, subtract, multiply, or divide. Watch the signs.

	a	b	c	d
9.	$52 + 6$	$38 - 5$	18×6	$9\overline{)54}$
10.	$76 - 12$	$8\overline{)35}$	$65 + 7$	50×7
11.	18×47	$63 + 24$	$9\overline{)40}$	$32 - 17$
12.	$6\overline{)168}$	47×56	$584 - 23$	$603 + 72$
13.	$578 + 49$	$372 - 68$	36×47	$7\overline{)861}$
14.	$705 - 183$	90×56	$8\overline{)3626}$	$897 + 268$

NAME ______________________

PRE-TEST—Addition and Subtraction

Chapter 1

Add.

	a	*b*	*c*	*d*	*e*	*f*
1.	2 +8	7 +5	9 +4	5 +5	6 +8	8 +9
2.	4 +7	8 +5	6 +4	9 +9	1 +9	8 +7
3.	8 +8	9 +5	6 +7	7 +3	4 +8	9 +3
4.	3 +8	6 +6	9 +2	7 +7	9 +7	6 +9

Subtract.

5.	10 − 3	12 − 8	15 − 6	14 − 5	18 − 9	16 − 8
6.	13 − 5	12 − 4	16 − 7	10 − 2	11 − 7	14 − 6
7.	10 − 5	12 − 6	11 − 2	15 − 7	17 − 9	10 − 8
8.	12 − 5	10 − 1	13 − 4	17 − 8	11 − 3	10 − 6
9.	12 − 9	15 − 8	16 − 9	11 − 5	13 − 6	14 − 7

Pre-Test—Problem Solving

Solve each problem.

1. Two adults and two children are playing. How many people are playing?

There are __________ adults.

There are __________ children.

There are __________ people playing.

1.

2. The Durhams played soccer for 1 hour and then played baseball for 2 hours. How many hours did they play in all?

They played soccer for __________ hour.

They played baseball for __________ hours.

They played __________ hours in all.

2.

3. The Durhams' house has 5 bedrooms in all. There are 2 bedrooms downstairs. The rest of the bedrooms are upstairs. How many bedrooms are upstairs?

There are __________ bedrooms in all.

There are __________ bedrooms downstairs.

There are __________ bedrooms upstairs.

3.

NAME ____________________

Lesson 1 Addition

$$\begin{array}{r} 2 \\ +6 \\ \hline 8 \end{array}$$

2 → Find the **2** -row.

+6 → Find the **6** -column.

8 ← The sum is named where the 2-row and 6-column meet.

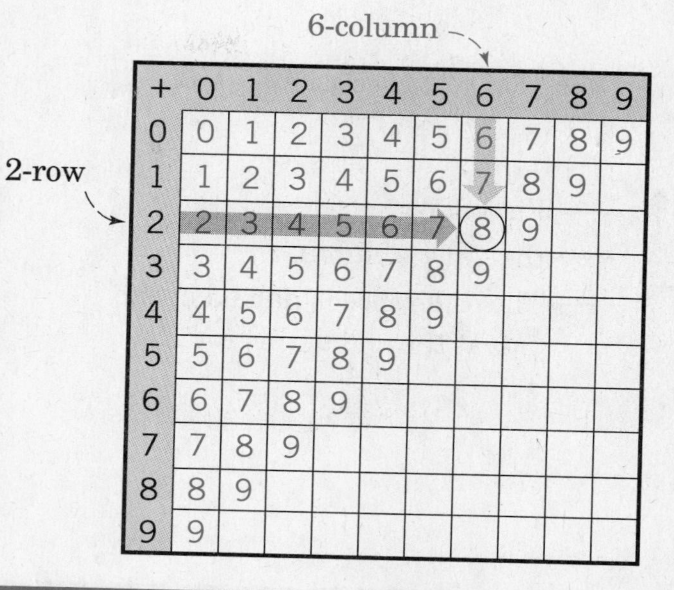

+	0	1	2	3	4	5	6	7	8	9
0	0	1	2	3	4	5	6	7	8	9
1	1	2	3	4	5	6	7	8	9	
2	2	3	4	5	6	7	8	9		
3	3	4	5	6	7	8	9			
4	4	5	6	7	8	9				
5	5	6	7	8	9					
6	6	7	8	9						
7	7	8	9							
8	8	9								
9	9									

Add.

	a	*b*	*c*	*d*	*e*	*f*	*g*	*h*
1.	2 +4	3 +1	1 +2	7 +0	0 +4	1 +4	5 +2	3 +3
2.	2 +0	6 +3	4 +4	3 +0	5 +3	1 +6	0 +5	8 +1
3.	2 +6	1 +0	1 +5	2 +2	3 +2	2 +1	5 +4	1 +7
4.	9 +0	5 +1	0 +3	4 +1	4 +5	1 +8	8 +0	4 +3
5.	0 +0	2 +3	7 +1	0 +9	4 +2	0 +2	0 +7	1 +1
6.	2 +7	0 +1	6 +2	0 +6	1 +3	6 +1	6 +0	7 +2

NAME ______________________

Lesson 2 Subtraction

6-column

−	0	1	2	3	4	5	6	7	8	9
0	0	1	2	3	4	5	6	7	8	9
1	1	2	3	4	5	6	7	8	9	
(2)	2	3	4	5	6	7	8	9		
3	3	4	5	6	7	8	9			
4	4	5	6	7	8	9				
5	5	6	7	8	9					
6	6	7	8	9						
7	7	8	9							
8	8	9								
9	9									

8 ⟶ Find 8 in
−6 ⟶ the 6 -column.
2 ⟵ The difference is named in the ▒ at the end of this row.

Subtract.

	a	*b*	*c*	*d*	*e*	*f*	*g*	*h*
1.	5 −4	3 −2	7 −7	1 −0	8 −2	9 −7	4 −3	6 −1
2.	7 −2	2 −2	7 −6	8 −7	9 −3	9 −8	4 −1	6 −0
3.	0 −0	7 −1	3 −0	6 −6	4 −2	6 −2	9 −5	8 −6
4.	9 −9	8 −4	9 −1	7 −5	7 −4	6 −5	2 −0	1 −1
5.	3 −1	9 −4	7 −3	5 −2	5 −1	6 −4	4 −4	8 −1
6.	5 −5	2 −1	5 −0	8 −3	9 −0	6 −3	7 −0	5 −3

NAME ____________________

Lesson 3 Addition

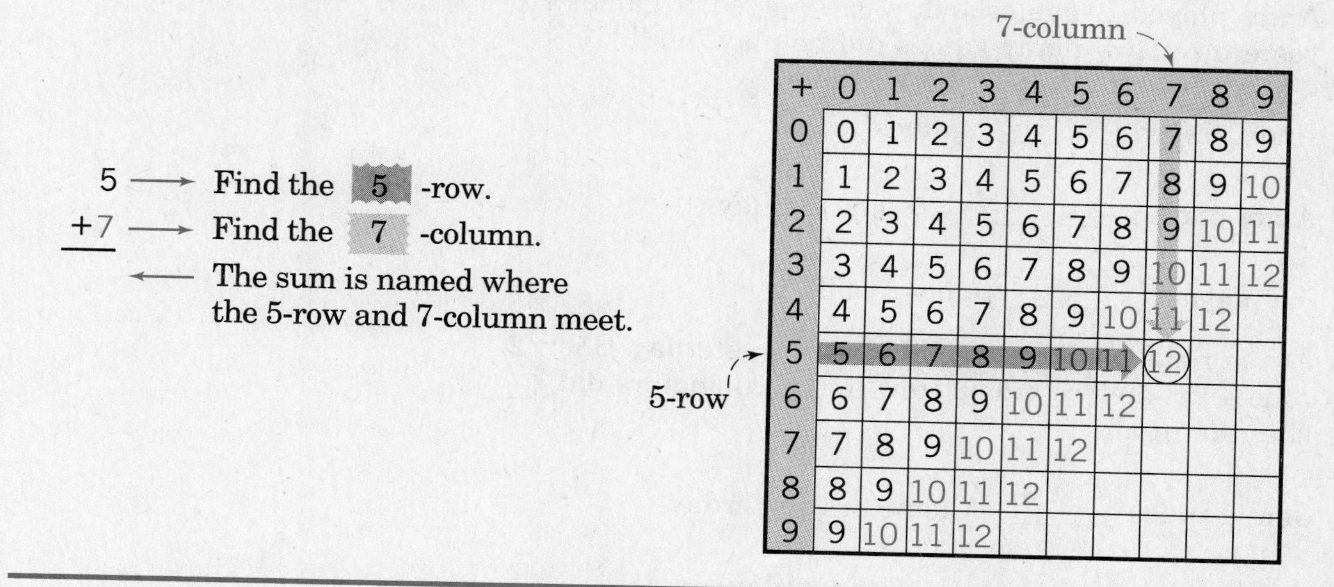

+	0	1	2	3	4	5	6	7	8	9
0	0	1	2	3	4	5	6	7	8	9
1	1	2	3	4	5	6	7	8	9	10
2	2	3	4	5	6	7	8	9	10	11
3	3	4	5	6	7	8	9	10	11	12
4	4	5	6	7	8	9	10	11	12	
5	5	6	7	8	9	10	11	12		
6	6	7	8	9	10	11	12			
7	7	8	9	10	11	12				
8	8	9	10	11	12					
9	9	10	11	12						

Add.

	a	*b*	*c*	*d*	*e*	*f*
1.	6 + 5	7 + 3	2 + 7	8 + 4	9 + 2	6 + 3
2.	8 + 2	3 + 9	3 + 5	5 + 2	6 + 4	5 + 5
3.	5 + 3	9 + 3	6 + 6	3 + 7	4 + 7	9 + 1
4.	5 + 7	8 + 1	5 + 6	2 + 8	2 + 5	7 + 5
5.	3 + 4	4 + 5	4 + 6	2 + 9	8 + 3	4 + 8
6.	2 + 6	1 + 9	3 + 8	7 + 1	7 + 4	6 + 2

Problem Solving

Solve each problem.

1. Andy played 2 games today. He played 9 games yesterday. How many games did he play in all?

Andy played ________ games today.

Andy played ________ games yesterday.

He played ________ games in all.

2. Jenna rode her bicycle 8 kilometers yesterday. She rode 4 kilometers today. How many kilometers did she ride in all?

Jenna rode ________ kilometers yesterday.

Jenna rode ________ kilometers today.

Jenna rode ________ kilometers in all.

3. Paul hit the ball 7 times. He missed 4 times. How many times did he swing at the ball?

Paul hit the ball ________ times.

Paul missed the ball ________ times.

Paul swung at the ball ________ times.

4. There were 4 people in a room. Six more people came in. How many people were in the room then?

________ people were in a room.

________ more people came in.

________ people were in the room then.

5. Heather and Justin each read 6 books. How many books did they read in all?

They read ________ books in all.

1.

2.

3.

4.

5.

NAME ____________________

Lesson 4 Subtraction

11 → Find 11 in
−4 → the 4 -column.
← The difference is named in the [shaded box] at the end of this row.

4-column

−	0	1	2	3	4	5	6	7	8	9
0	0	1	2	3	4	5	6	7	8	9
1	1	2	3	4	5	6	7	8	9	10
2	2	3	4	5	6	7	8	9	10	11
3	3	4	5	6	7	8	9	10	11	12
4	4	5	6	7	8	9	10	11	12	
5	5	6	7	8	9	10	11	12		
6	6	7	8	9	10	11	12			
(7)	7	8	9	10	11	12				
8	8	9	10	11	12					
9	9	10	11	12						

Subtract.

	a	*b*	*c*	*d*	*e*	*f*
1.	11 −7	10 −4	10 −8	12 −9	8 −5	11 −2
2.	10 −1	11 −8	7 −4	11 −6	12 −3	9 −6
3.	12 −7	10 −7	9 −3	11 −9	12 −4	10 −5
4.	8 −6	12 −8	9 −5	10 −6	11 −5	8 −8
5.	12 −6	10 −9	9 −8	7 −6	11 −4	9 −7
6.	10 −2	7 −3	10 −3	12 −5	8 −3	11 −3

Problem Solving

Solve each problem.

1. There were 12 nails in a box. David used 3 of them. How many nails are still in the box?

   ________ nails were in a box.

   ________ nails were used.

   ________ nails are still in the box.

2. There are 11 checkers on a board. Eight of them are black. The rest are red. How many red checkers are on the board?

   ________ checkers are on a board.

   ________ checkers are black, and the rest are red.

   ________ red checkers are on the board.

3. Marty is 10 years old. Her brother Larry is 7. Marty is how many years older than Larry?

   Marty's age is ________ years.

   Larry's age is ________ years.

   Marty is ________ years older than Larry.

4. Joye walked 11 blocks. Ann walked 2 blocks. Joye walked how much farther than Ann?

   Joye walked ________ blocks.

   Ann walked ________ blocks.

   Joye walked ________ blocks farther than Ann.

5. Twelve people are in a room. Five of them are men. How many are women?

   ________ women are in the room.

1.

2.

3.

4.

5.

NAME ____________________

Lesson 5 Addition and Subtraction

To check 5 + 6 = 11, subtract 6 from 11.

$$\begin{array}{r} 5 \\ +6 \\ \hline 11 \\ -6 \\ \hline 5 \end{array}$$

These should be the same.

To check 13 − 4 = 9, add 4 to ________.

$$\begin{array}{r} 13 \\ -4 \\ \hline 9 \\ +4 \\ \hline 13 \end{array}$$

These should be the same.

Add. Check each answer.

	a	*b*	*c*	*d*	*e*	*f*
1.	$2 + 9$	$8 + 4$	$7 + 3$	$3 + 8$	$1 + 9$	$6 + 6$
2.	$9 + 3$	$5 + 6$	$4 + 8$	$5 + 5$	$7 + 4$	$9 + 1$

Subtract. Check each answer.

3.	$10 - 8$	$12 - 7$	$11 - 3$	$10 - 4$	$11 - 7$	$10 - 7$
4.	$11 - 9$	$12 - 8$	$11 - 8$	$12 - 5$	$10 - 6$	$10 - 3$

Problem Solving

Answer each question.

1. Ben had some marbles. He gave 2 of them away and had 9 left. How many marbles did he start with?

   Are you to add or subtract? ______________

   How many marbles did he start with? __________

   1.

2. A full box has 10 pieces of chalk. This box has only 8 pieces. How many pieces are missing?

   Are you to add or subtract? ______________

   How many pieces are missing? ____________

   2.

3. Noah is 11 years old today. How old was he 4 years ago?

   Are you to add or subtract? ______________

   How old was Noah 4 years ago? ____________

   3.

4. Nine boys were playing ball. Then 3 more boys began to play. How many boys were playing ball then?

   Are you to add or subtract? ______________

   How many boys were playing then? __________

   4.

5. Alyssa has as many sisters as brothers. She has 5 brothers. How many brothers and sisters does she have?

   Are you to add or subtract? ______________

   How many brothers and sisters does Alyssa have? ______________

   5.

6. Tricia invited 12 people to her party. Seven people came. How many people that were invited did not come?

   Are you to add or subtract? ______________

   How many people did not come? ____________

   6.

Lesson 6 Addition

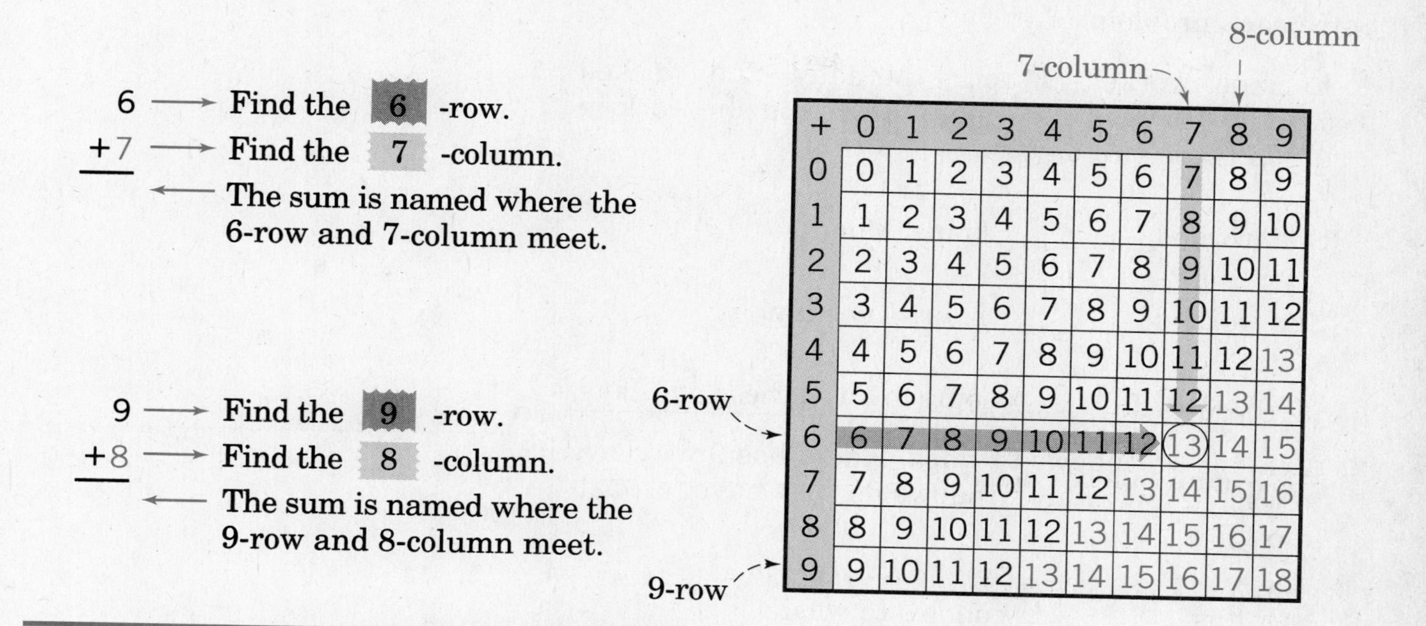

6 ⟶ Find the 6 -row.
+7 ⟶ Find the 7 -column.
⟵ The sum is named where the 6-row and 7-column meet.

9 ⟶ Find the 9 -row.
+8 ⟶ Find the 8 -column.
⟵ The sum is named where the 9-row and 8-column meet.

+	0	1	2	3	4	5	6	7	8	9
0	0	1	2	3	4	5	6	7	8	9
1	1	2	3	4	5	6	7	8	9	10
2	2	3	4	5	6	7	8	9	10	11
3	3	4	5	6	7	8	9	10	11	12
4	4	5	6	7	8	9	10	11	12	13
5	5	6	7	8	9	10	11	12	13	14
6	6	7	8	9	10	11	12	13	14	15
7	7	8	9	10	11	12	13	14	15	16
8	8	9	10	11	12	13	14	15	16	17
9	9	10	11	12	13	14	15	16	17	18

Add.

	a	*b*	*c*	*d*	*e*	*f*
1.	7 +6	8 +7	7 +4	9 +7	4 +9	8 +8
2.	5 +9	6 +4	6 +8	5 +8	8 +4	7 +8
3.	6 +9	5 +5	6 +7	9 +2	8 +6	4 +6
4.	5 +7	8 +9	9 +6	5 +6	9 +4	9 +9
5.	7 +9	8 +2	9 +8	8 +5	9 +1	4 +7
6.	9 +5	6 +6	2 +9	4 +8	7 +7	9 +3

Problem Solving

Solve each problem.

1. Luciano worked 9 hours Monday. She worked 7 hours Tuesday. How many hours did she work in all on those two days?

She worked ________ hours Monday.

She worked ________ hours Tuesday.

She worked ________ hours in all on those two days.

2. Alex has 6 windows to wash. Nadia has 9 windows to wash. How many windows do they have to wash in all?

Alex has ________ windows to wash.

Nadia has ________ windows to wash.

Together they have ________ windows to wash.

3. Seven cars are in the first row. Six cars are in the second row. How many cars are in the first two rows?

________ cars are in the first two rows.

4. There are 9 men and 8 women at work. How many people are at work?

There are ________ people at work.

5. Andrew worked 8 hours. Geraldo worked the same number of hours. How many hours did Andrew and Geraldo work in all?

They worked ________ hours in all.

6. There are 6 plants in a box. Seven more are on a table. How many plants are there?

There are ________ plants.

1.	
2.	
3.	4.
5.	6.

NAME ____________________

Lesson 7 Subtraction

13 → Find 13 in
−8 → the 8 -column.
← The difference is named in the ▒ at the end of that row.

15 → Find 15 in
−6 → the 6 -column.
← The difference is named in the ▒ at the end of that row.

6-column ↘ 8-column ↓

−	0	1	2	3	4	5	6	7	8	9
0	0	1	2	3	4	5	6	7	8	9
1	1	2	3	4	5	6	7	8	9	10
2	2	3	4	5	6	7	8	9	10	11
3	3	4	5	6	7	8	9	10	11	12
4	4	5	6	7	8	9	10	11	12	13
(5)	5	6	7	8	9	10	11	12	13	14
6	6	7	8	9	10	11	12	13	14	15
7	7	8	9	10	11	12	13	14	15	16
8	8	9	10	11	12	13	14	15	16	17
9	9	10	11	12	13	14	15	16	17	18

Subtract.

	a	*b*	*c*	*d*	*e*	*f*
1.	13 −5	14 −8	16 −7	10 −9	12 −5	14 −6
2.	17 −8	13 −7	12 −4	14 −5	15 −8	13 −6
3.	11 −7	18 −9	15 −6	11 −8	14 −7	13 −9
4.	16 −8	10 −5	12 −7	13 −4	12 −6	14 −9
5.	13 −8	12 −9	10 −1	15 −9	11 −3	10 −7
6.	15 −7	10 −3	17 −9	11 −6	16 −9	11 −4

Problem Solving

Solve each problem.

1. Matt wants to collect 13 cars. He now has 5 cars. How many more cars does he need?

   Matt wants ________ cars.

   He now has ________ cars.

   He needs ________ more cars.

2. Susan bought 18 valentines. She mailed 9 of them. How many valentines does she have left?

   Susan bought ________ valentines.

   She mailed ________ of them.

   She has ________ valentines left.

3. Courtney had 16 stamps. She used some, and had 7 left. How many stamps did she use?

   Courtney used ________ stamps.

4. Bret is 14 years old. Amy is 7. Bret is how much older than Amy?

   Bret is ________ years older than Amy.

5. Fifteen bolts and nuts were on the table. Seven were bolts. How many were nuts?

   There were ________ nuts.

6. There are 17 machine parts in a drawer. Only nine are new parts. How many are not new parts?

   ________ are not new parts.

1.

2.

3.

4.

5.

6.

NAME ____________________

Lesson 8 Addition and Subtraction

To check 6 + 8 = 14, subtract 8 from 14.	6 + 8 14 −8 6	These should be the same.	To check 13 − 6 = 7, add ________ to 7.	13 − 6 7 +6 13	These should be the same.

Add. Check each answer.

	a	*b*	*c*	*d*	*e*	*f*
1.	5 +9	9 +7	6 +6	7 +4	9 +8	3 +7
2.	6 +7	9 +3	6 +9	4 +9	6 +4	8 +6

Subtract. Check each answer.

3.	14 −8	18 −9	13 −5	15 −6	16 −8	12 −7
4.	13 −6	12 −4	13 −4	16 −9	15 −7	13 −8

Problem Solving

Answer each problem.

1. Penny worked 9 addition problems. She worked 7 subtraction problems. How many problems did she work?

Are you to add or subtract? ____________

How many problems did she work? ____________

2. Six people were in the room. Then 8 more people came in. How many people were in the room then?

Are you to add or subtract? ____________

How many people were in the room then? ________

3. There were 18 chairs in a room. Nine of them were being used. How many were not being used?

Are you to add or subtract? ____________

How many chairs were not being used? ____________

4. Mr. Noe and Miss Leikel had 17 students absent. Mr. Noe had 9 absent. How many did Miss Leikel have absent?

Are you to add or subtract? ____________

How many students were absent from Miss Leikel's class? ____________

5. There were 14 children at the park. Five were boys. How many were girls?

Are you to add or subtract? ____________

How many girls were at the park? ____________

1.

2.

3.

4.

5.

NAME ____________________

CHAPTER 1 TEST

Add.

	a	b	c	d	e	f
1.	7 +8	4 +9	6 +5	2 +8	8 +6	7 +5
2.	9 +1	5 +8	8 +4	9 +2	8 +8	5 +9
3.	5 +7	6 +9	8 +3	8 +9	3 +8	9 +4
4.	9 +6	6 +7	9 +9	6 +6	7 +9	6 +4
5.	8 +5	3 +9	1 +9	7 +4	3 +7	6 +8

Subtract.

	a	b	c	d	e
6.	10 −6	14 −7	12 −3	15 −7	12 −8
7.	13 −4	16 −7	11 −9	10 −5	13 −6
8.	10 −2	14 −6	11 −6	12 −7	17 −9
9.	16 −8	11 −7	10 −3	14 −8	15 −8

NAME ____________________

PRE-TEST—Addition and Subtraction

Chapter 2

Add.

	a	b	c	d	e	f
1.	3 +6	43 +6	1 +4	51 +4	2 +5	82 +5
2.	57 +2	26 +1	44 +3	23 +4	42 +3	21 +5
3.	4 +31	5 +43	4 +62	3 +43	5 +12	7 +20
4.	54 +31	26 +12	45 +33	67 +21	42 +33	22 +13
5.	32 +31	24 +24	36 +61	20 +19	45 +23	32 +21

Subtract.

	a	b	c	d	e	f
6.	7 −4	37 − 4	5 −2	45 −2	8 −6	38 −6
7.	38 −4	27 −6	54 −3	29 −7	68 −2	26 − 3
8.	54 −23	69 −24	37 −21	88 −24	93 −21	87 −37
9.	28 −13	54 −34	87 −26	54 −21	50 −40	37 −10

NAME ______________________

Lesson 1 Addition

	Add the ones.	Add the tens.
36 + 2	36 + 2 = 8	36 + 2 = 38
6 + 41	6 + 41 = 7	6 + 41 = 47

Add.

	a	*b*	*c*	*d*	*e*	*f*
1.	3 + 5	23 + 5	2 + 3	42 + 3	5 + 1	25 + 1
2.	3 + 4	3 + 64	4 + 5	4 + 55	2 + 5	2 + 85
3.	2 + 4	12 + 4	22 + 4	32 + 4	42 + 4	52 + 4
4.	5 + 63	6 + 31	24 + 3	92 + 2	57 + 1	2 + 41
5.	41 + 3	21 + 2	3 + 63	2 + 84	21 + 6	4 + 14
6.	8 + 51	62 + 4	25 + 3	6 + 33	2 + 51	5 + 43
7.	36 + 2	42 + 5	2 + 51	60 + 8	5 + 21	34 + 2

Problem Solving

Solve each problem.

1. John has 32 red marbles and 5 green marbles. How many red and green marbles does he have? John has __________ red and green marbles.	**1.**
2. Su-Lee had 5 paper cups. She bought 24 more. How many paper cups did she have then? She then had __________ paper cups.	**2.**
3. On the way to work, Michael counted 41 cars and 7 trucks. How many cars and trucks did he count? Michael counted __________ cars and trucks.	**3.**
4. Mark worked all the problems on a test. He had 24 right answers and 4 wrong ones. How many problems were on the test? There were __________ problems on the test.	**4.**
5. Shea works with 12 women and 6 men. How many people does she work with? Shea works with __________ people.	**5.**
6. Four men and 11 women are on the bus. How many people are on the bus? __________ people are on the bus.	**6.**
7. Marta weighs 20 kilograms. Her baby brother weighs 4 kilograms. How much do they weigh together? They weigh __________ kilograms together.	**7.**

NAME ______________________

Lesson 2 Addition

	Add the ones.	Add the tens.	
36 +43	36 +43 9	36 +43 79	25 +61 86

86: 6 ← Add the ones. 8 ← Add the tens.

Add.

	a	b	c	d	e	f
1.	23 +45	63 +21	45 +22	61 +30	42 +35	60 +25
2.	48 +41	52 +14	32 +54	63 +20	21 +38	45 +52
3.	34 +22	41 +25	36 +22	51 +40	83 +12	42 +30
4.	63 +24	30 +58	27 +12	44 +23	62 +14	35 +53
5.	24 +31	52 +32	42 +27	51 +33	16 +20	43 +23
6.	34 +25	64 +23	18 +41	54 +24	41 +27	14 +32

Problem Solving

Solve each problem.

1. There are 12 boys and 13 girls in Jean's class. How many students are in her class?

   There are ________ students in her class.

2. Emily scored 32 baskets. She missed 23 times. How many times did she try to score?

   Emily tried to score ________ times.

3. One store ordered 52 bicycles. Another store ordered 45 bicycles. How many bicycles did both stores order?

   Both stores ordered ________ bicycles.

4. One bear cub weighs 64 kilograms. Another bear cub is 22 kilograms heavier. How much does the heavier cub weigh?

   The heavier cub weighs________ kilograms.

5. Joshua rode the bus 42 blocks east and 25 blocks south. How many blocks did Joshua ride the bus?

   Joshua rode the bus ________ blocks.

6. 43 women and 35 men came to the meeting. How many people came to the meeting?

   ________ people came to the meeting.

7. 68 seats were filled, and 21 were empty. How many seats were there?

   There were ________ seats.

1.	
2.	3.
4.	5.
6.	7.

NAME ____________________

Lesson 3 Subtraction

	Subtract the ones.	**Subtract the tens.**		Subtract the ones.	**Subtract the tens.**
47	47	47	64	64	64
−2	−2	−2	−23	−23	−23
	5	45		1	41

Subtract.

	a	*b*	*c*	*d*	*e*	*f*
1.	9 − 3	49 − 3	5 − 2	35 − 2	7 − 1	87 − 1
2.	8 − 2	78 − 2	4 − 3	64 − 3	9 − 9	89 − 9
3.	45 − 3	36 − 4	78 − 5	42 − 2	38 − 8	65 − 4
4.	49 − 26	37 − 16	58 − 23	49 − 31	78 − 45	73 − 20
5.	58 − 27	69 − 31	42 − 21	49 − 19	84 − 23	78 − 64
6.	78 − 21	67 − 31	40 − 20	56 − 36	45 − 23	92 − 21
7.	56 − 41	85 − 63	94 − 32	77 − 46	99 − 32	86 − 23

Problem Solving

Solve each problem.

1. Beth worked 27 problems. She got 6 wrong answers. How many answers did she get right?

Beth got __________ answers right.

2. There were 96 parts in a box. Four parts were broken. How many parts were not broken?

______________ parts were not broken.

3. At noon the temperature was 28 degrees Celsius. At nine o'clock in the evening, it was 14 degrees Celsius. How many degrees did the temperature drop?

The temperature dropped _____________ degrees.

4. Clark had 75 cents. Then he spent 25 cents for some paper. How many cents did he have left?

Clark had____________ cents left.

5. There are 72 houses in Kyle's neighborhood. Kyle delivers papers to all but 21 of them. How many houses does he deliver papers to?

He delivers papers to ____________ houses.

6. Ninety-five students were in the gym. Thirty-four were boys. How many were girls?

_______________ students were girls.

7. A rope is 47 inches long. A wire is 17 inches long. How much longer is the rope?

The rope is _______________ inches longer.

1.	
2.	**3.**
4.	**5.**
6.	**7.**

NAME ______________________

Lesson 4 Subtraction

To check
37 − 24 = 13,
add 24
to ________.

$$\begin{array}{r} 37 \\ -24 \\ \hline 13 \\ +24 \\ \hline 37 \end{array}$$

These should be the same.

To check
59 − 29 = 30,
add ________
to 30.

$$\begin{array}{r} 59 \\ -29 \\ \hline 30 \\ +29 \\ \hline 59 \end{array}$$

These should be the same.

Subtract. Check each answer.

	a	*b*	*c*	*d*	*e*	*f*
1.	$\begin{array}{r} 59 \\ -34 \\ \hline \end{array}$	$\begin{array}{r} 27 \\ -14 \\ \hline \end{array}$	$\begin{array}{r} 85 \\ -23 \\ \hline \end{array}$	$\begin{array}{r} 78 \\ -23 \\ \hline \end{array}$	$\begin{array}{r} 47 \\ -24 \\ \hline \end{array}$	$\begin{array}{r} 59 \\ -26 \\ \hline \end{array}$
2.	$\begin{array}{r} 85 \\ -25 \\ \hline \end{array}$	$\begin{array}{r} 48 \\ -32 \\ \hline \end{array}$	$\begin{array}{r} 56 \\ -24 \\ \hline \end{array}$	$\begin{array}{r} 96 \\ -35 \\ \hline \end{array}$	$\begin{array}{r} 40 \\ -30 \\ \hline \end{array}$	$\begin{array}{r} 92 \\ -81 \\ \hline \end{array}$
3.	$\begin{array}{r} 74 \\ -23 \\ \hline \end{array}$	$\begin{array}{r} 58 \\ -26 \\ \hline \end{array}$	$\begin{array}{r} 75 \\ -24 \\ \hline \end{array}$	$\begin{array}{r} 38 \\ -23 \\ \hline \end{array}$	$\begin{array}{r} 45 \\ -35 \\ \hline \end{array}$	$\begin{array}{r} 88 \\ -35 \\ \hline \end{array}$
4.	$\begin{array}{r} 67 \\ -24 \\ \hline \end{array}$	$\begin{array}{r} 87 \\ -24 \\ \hline \end{array}$	$\begin{array}{r} 59 \\ -36 \\ \hline \end{array}$	$\begin{array}{r} 58 \\ -24 \\ \hline \end{array}$	$\begin{array}{r} 79 \\ -54 \\ \hline \end{array}$	$\begin{array}{r} 84 \\ -23 \\ \hline \end{array}$

Problem Solving

Solve each problem.

1. Mr. Ming wants to build a fence 58 meters long. He has 27 meters of fence completed. How much of the fence is left to build?

   ________ meters of fence is left to build.

2. Mrs. Boyle is taking an 89-mile trip. She has traveled 64 miles. How much farther must she travel?

   Mrs. Boyle must travel ________ more miles.

3. Sean had 95 cents. Then he spent 45 cents. How many cents did he have left?

   Sean had ________ cents left.

4. Kevin scored 62 points and Bianca scored 78 points. How many more points did Bianca score than Kevin?

   Bianca scored ________ more points.

5. Darien lives 38 blocks from the ball park. Kelly lives 25 blocks from the park. How much farther from the ball park does Darien live than Kelly?

   Darien lives ________ blocks farther than Kelly.

6. Eighty-four students were in the pool. Fifty of them were boys. How many girls were in the pool?

   ________ girls were in the pool.

7. Chad said that 88 buses stop at Division Street each day. So far, 13 buses have stopped. How many more buses should stop today?

   ________ more buses should stop.

1.	
2.	3.
4.	5.
6.	7.

NAME ______________________

Lesson 5 Addition and Subtraction

To check 43 + 14 = 57, subtract 14 from ________.

```
  43
+ 14
  57
- 14
  43
```

These should be the same.

To check 57 − 14 = 43, add ________ to 43.

```
  57
- 14
  43
+ 14
  57
```

These should be the same.

Add. Check each answer.

	a	*b*	*c*	*d*	*e*	*f*
1.	27 + 31	42 + 51	26 + 30	14 + 52	23 + 72	65 + 22
2.	44 + 24	31 + 27	64 + 14	32 + 20	42 + 36	46 + 23

Subtract. Check each answer.

3.	78 − 23	48 − 13	27 − 16	58 − 26	67 − 24	38 − 16
4.	75 − 61	46 − 26	39 − 10	45 − 23	67 − 41	38 − 15

Problem Solving

Solve each problem.

1. Mrs. Dial weighs 55 kilograms. Her son weighs 32 kilograms. How much more than her son does Mrs. Dial weigh?

   She weighs ________ kilograms more.

2. Mitzi planted 55 flower seeds. Only 23 of them grew. How many did not grow?

   ________ seeds did not grow.

3. A city has 48 mail trucks. Twelve are not being used today. How many mail trucks are being used?

   ________ mail trucks are being used.

4. A mail carrier delivered 38 letters and picked up 15. How many more letters were delivered than were picked up?

   The carrier delivered ________ more letters.

5. A city has 89 mail carriers. One day 77 carriers were at work. How many were not at work?

   ________ carriers were not at work.

6. Our mail carrier walks about 32 miles each week. About how many miles does our carrier walk in two weeks?

   Our carrier walks ________ miles in two weeks.

7. Ms. Tottle worked at a store for 23 years. She then worked 26 years at a bank. How many years did she work at these two places?

   She worked ________ years at these two places.

1.	
2.	3.
4.	5.
6.	7.

NAME ____________________

CHAPTER 2 TEST

Add or subtract. Check each answer.

	a	*b*	*c*	*d*	*e*	*f*
1.	$\begin{array}{r} 4 \\ +2 \\ \hline \end{array}$	$\begin{array}{r} 24 \\ +2 \\ \hline \end{array}$	$\begin{array}{r} 36 \\ +3 \\ \hline \end{array}$	$\begin{array}{r} 4 \\ +21 \\ \hline \end{array}$	$\begin{array}{r} 72 \\ +4 \\ \hline \end{array}$	$\begin{array}{r} 9 \\ +30 \\ \hline \end{array}$
2.	$\begin{array}{r} 36 \\ +21 \\ \hline \end{array}$	$\begin{array}{r} 41 \\ +38 \\ \hline \end{array}$	$\begin{array}{r} 65 \\ +22 \\ \hline \end{array}$	$\begin{array}{r} 41 \\ +26 \\ \hline \end{array}$	$\begin{array}{r} 35 \\ +50 \\ \hline \end{array}$	$\begin{array}{r} 66 \\ +21 \\ \hline \end{array}$
3.	$\begin{array}{r} 7 \\ -2 \\ \hline \end{array}$	$\begin{array}{r} 37 \\ -2 \\ \hline \end{array}$	$\begin{array}{r} 45 \\ -4 \\ \hline \end{array}$	$\begin{array}{r} 26 \\ -3 \\ \hline \end{array}$	$\begin{array}{r} 49 \\ -8 \\ \hline \end{array}$	$\begin{array}{r} 27 \\ -5 \\ \hline \end{array}$
4.	$\begin{array}{r} 48 \\ -23 \\ \hline \end{array}$	$\begin{array}{r} 65 \\ -24 \\ \hline \end{array}$	$\begin{array}{r} 45 \\ -22 \\ \hline \end{array}$	$\begin{array}{r} 68 \\ -28 \\ \hline \end{array}$	$\begin{array}{r} 54 \\ -22 \\ \hline \end{array}$	$\begin{array}{r} 67 \\ -30 \\ \hline \end{array}$

Solve.

5. Miss Jones has 32 students. Mr. Lum has 26 students. How many students are in the two classes?

There are __________ students in the two classes.

NAME ____________________

Chapter 3

PRE-TEST—Addition and Subtraction

Add.

	a	*b*	*c*	*d*	*e*	*f*
1.	53 +6	24 +2	2 +35	8 +81	64 +3	25 +2
2.	36 +5	54 +8	8 +39	2 +59	48 +8	26 +7
3.	42 +33	72 +14	54 +23	61 +28	19 +40	26 +52
4.	54 +27	35 +36	59 +38	54 +19	27 +48	39 +39
5.	49 +23	62 +17	43 +21	48 +48	26 +40	56 +37

Subtract.

	a	*b*	*c*	*d*	*e*	*f*
6.	37 −3	29 −4	54 −4	87 −2	56 −5	89 −6
7.	47 −9	72 −5	45 −7	55 −9	40 −5	34 −7
8.	54 −12	42 −30	75 −64	46 −23	93 −81	89 −41
9.	73 −25	85 −49	92 −24	64 −56	77 −48	88 −38

NAME ______________________________

Lesson 1 Addition

Add the ones.
Rename 13 as 10 + 3.

Add the tens.

```
 54       4           1        1
+ 9      +9          54       54
        13 or 10 + 3 + 9      + 9
                      3       63
```

Add.

	a	b	c	d	e	f
1.	27 + 5	35 + 8	87 + 4	38 + 9	42 + 8	46 + 5
2.	45 + 9	27 + 7	7 + 38	20 + 65	24 + 9	8 + 38
3.	27 + 3	45 + 6	8 + 36	9 + 29	6 + 58	42 + 9
4.	76 + 7	3 + 47	4 + 26	27 + 4	5 + 18	9 + 19
5.	6 + 15	41 + 9	52 + 8	65 + 9	7 + 38	6 + 16
6.	9 + 28	36 + 7	59 + 2	7 + 36	4 + 47	9 + 38
7.	46 + 8	9 + 25	8 + 68	4 + 59	85 + 5	78 + 7

Problem Solving

Solve each problem.

1. Last year there were 44 monkeys on an island. There are 8 more monkeys this year. How many monkeys are on the island now?

   There were ________ monkeys last year.

   There are ________ more monkeys this year.

   There are ________ monkeys on the island now.

2. There were 72 children and 9 adults in our group at the zoo. How many people were in our group?

   ________ children were in our group.

   ________ adults were in our group.

   ________ people were in our group.

3. One group of monkeys was fed 6 kilograms of fruit. Another group was fed 19 kilograms. How much fruit was that in all?

   That was ________ kilograms of fruit in all.

4. The children drank 68 cartons of milk. There were 8 cartons left. How many cartons of milk were there to start with?

   There were ________ cartons of milk to start with.

5. A zoo has 87 kinds of snakes. They are getting 4 new kinds. How many kinds will they have then?

   Then they will have ________ kinds of snakes.

1.	
2.	
3.	
4.	5.

NAME ____________________

Lesson 2 Addition

Add the ones.
Rename 15 as 10 + 5.

Add the tens.

$$\begin{array}{r} 48 \\ +27 \\ \hline \end{array} \qquad \begin{array}{r} 8 \\ +7 \\ \hline 15 \end{array} \text{ or } 10 + 5 \qquad \begin{array}{r} {}^{1} \\ 48 \\ +27 \\ \hline 5 \end{array} \qquad \begin{array}{r} {}^{1} \\ 48 \\ +27 \\ \hline 75 \end{array}$$

Add.

	a	*b*	*c*	*d*	*e*	*f*
1.	37 + 25	48 + 37	26 + 54	35 + 29	54 + 18	62 + 29
2.	29 + 28	38 + 37	47 + 25	63 + 27	79 + 19	64 + 17
3.	58 + 26	45 + 18	27 + 57	44 + 29	36 + 36	77 + 17
4.	49 + 48	26 + 37	73 + 19	18 + 28	15 + 47	29 + 27
5.	18 + 55	28 + 24	38 + 37	48 + 43	58 + 16	68 + 28
6.	26 + 66	19 + 54	57 + 29	45 + 36	52 + 18	33 + 29
7.	35 + 56	47 + 28	31 + 39	29 + 59	67 + 16	55 + 28

Problem Solving

Solve each problem.

1. January has 31 days. February has 29 days this year. How many days are in the two months?

There are ________ days in January.

There are ________ days in February this year.

There are ________ days in January and February.

1.

2. Jeff weighs 46 kilograms. His father is 36 kilograms heavier. How much does Jeff's father weigh?

Jeff weighs ________ kilograms.

His father is ________ kilograms heavier.

His father weighs ________ kilograms.

2.

3. Lauren had 29 points. She earned 13 more. How many points did she have then?

Lauren had ________ points.

She earned ________ more.

She had ________ points then.

3.

4. Adam gained 18 pounds in the last two years. Two years ago he weighed 59 pounds. How much does he weigh today?

Adam weighs ________ pounds today.

5. Kathy read 25 pages of a story. She has 36 more pages to read. How many pages are there in the story?

There are ________ pages in the story.

4. **5.**

NAME ____________________

Lesson 3 Subtraction

	To subtract the ones, rename 63 as "5 tens and 13 ones."	Subtract the ones.	Subtract the tens.
$\begin{array}{r} 63 \\ -9 \\ \hline \end{array}$	$\begin{array}{r} \scriptstyle 5\,13 \\ \not{6}\,\not{3} \\ -9 \\ \hline \end{array}$	$\begin{array}{r} \scriptstyle 5\,13 \\ \not{6}\,\not{3} \\ -9 \\ \hline 4 \end{array}$	$\begin{array}{r} \scriptstyle 5\,13 \\ \not{6}\,\not{3} \\ -9 \\ \hline 54 \end{array}$

Subtract.

	a	*b*	*c*	*d*	*e*	*f*
1.	53 − 8	27 − 9	46 − 9	54 − 5	32 − 6	65 − 7
2.	28 − 9	48 − 9	35 − 6	44 − 7	67 − 8	92 − 9
3.	52 − 6	62 − 4	61 − 6	73 − 5	50 − 9	42 − 5
4.	96 − 8	73 − 6	80 − 7	42 − 3	63 − 4	51 − 9
5.	94 − 8	88 − 9	33 − 4	27 − 9	46 − 8	64 − 7
6.	23 − 9	76 − 8	40 − 4	41 − 6	53 − 7	25 − 7
7.	47 − 8	31 − 7	82 − 8	74 − 6	93 − 9	60 − 5

Problem Solving

Solve each problem.

1. There were 48 words on a spelling test. Sarah missed 9 of them. How many words did she spell correctly?

There were ________ words on the test.

Sarah missed ________ words.

She spelled ________ words correctly.

2. Ryan earned 91 points. Mike earned 5 points less than Ryan. How many points did Mike earn?

Ryan earned ________ points.

Mike earned ________ points less than Ryan.

Mike earned ________ points.

3. Sheila lost 7 of the 45 games she played. How many games did she win?

She won ________ games.

4. Travis had 50 tickets to sell. He sold some and had 6 left. How many tickets did he sell?

Travis sold ________ tickets.

5. There were 73 books in the classroom library. Some of the books are checked out. Seven are still there. How many books are checked out?

________ books are checked out.

6. Angela's great-grandfather is 82 years old. How old was he 4 years ago?

Four years ago he was________ years old.

1.

2.

3. **4.**

5. **6.**

NAME ______________________

Lesson 4 Subtraction

To subtract the ones, rename 92 as "8 tens and 12 ones."

Subtract the ones.

Subtract the tens.

$$\begin{array}{r} 92 \\ -38 \\ \hline \end{array} \qquad \begin{array}{r} {\scriptstyle 8\,12} \\ \not{9}\not{2} \\ -3\,8 \\ \hline \end{array} \qquad \begin{array}{r} {\scriptstyle 8\,12} \\ \not{9}\not{2} \\ -3\,8 \\ \hline 4 \end{array} \qquad \begin{array}{r} {\scriptstyle 8\,12} \\ \not{9}\not{2} \\ -3\,8 \\ \hline 54 \end{array}$$

Subtract.

	a	*b*	*c*	*d*	*e*	*f*
1.	35 − 17	27 − 19	54 − 37	63 − 26	84 − 59	28 − 19
2.	42 − 24	56 − 39	41 − 27	53 − 15	86 − 78	92 − 26
3.	43 − 15	37 − 29	26 − 19	55 − 36	43 − 27	28 − 19
4.	54 − 26	35 − 18	22 − 15	56 − 29	38 − 19	31 − 18
5.	83 − 25	94 − 16	65 − 39	73 − 17	80 − 28	92 − 35
6.	35 − 26	90 − 55	56 − 27	41 − 16	50 − 38	61 − 15
7.	52 − 18	75 − 38	47 − 39	60 − 11	86 − 59	94 − 48

Problem Solving

Solve each problem.

1. Joseph weighs 95 pounds. Zach weighs 26 pounds less than Joseph. How much does Zach weigh?

Joseph weighs ________ pounds.

Zach weighs ________ pounds less than Joseph.

Zach weighs ________ pounds.

2. There are 73 children in the gym. Forty-five of them are boys. How many girls are in the gym?

There are ________ children in the gym.

There are ________ boys in the gym.

There are ________ girls in the gym.

3. A store has 84 bicycles. They have 45 girls' bicycles. How many boys' bicycles do they have?

________ bicycles are boys' bicycles.

4. It takes 50 points to win a prize. Paige has 38 points. How many more points does Paige need to win a prize?

Paige needs ________ points.

5. Allison has 19 more pages to read in a book. The book has 46 pages in all. How many pages has Allison already read?

Allison has already read ________ pages.

6. The Tigers scored 33 points. The Bears scored 18 points. How many more points did the Tigers score than the Bears?

The Tigers scored ________ more points.

1.	
2.	
3.	**4.**
5.	**6.**

NAME ____________________

Lesson 5 Addition and Subtraction

To check 34 + 19 = 53, subtract 19 from ________.

$$\begin{array}{r} 34 \\ +19 \\ \hline 53 \\ -19 \\ \hline 34 \end{array}$$

These should be the same.

To check 53 − 19 = 34, add ________ to 34.

$$\begin{array}{r} 53 \\ -19 \\ \hline 34 \\ +19 \\ \hline 53 \end{array}$$

These should be the same.

Add. Check each answer.

	a	*b*	*c*	*d*	*e*	*f*
1.	54 + 7	46 + 9	63 + 18	58 + 27	21 + 49	45 + 46
2.	26 + 38	37 + 19	41 + 9	58 + 18	67 + 27	35 + 38

Subtract. Check each answer.

	a	*b*	*c*	*d*	*e*	*f*
3.	62 − 8	48 − 9	35 − 16	96 − 29	52 − 14	43 − 5
4.	36 − 18	57 − 8	67 − 19	52 − 17	51 − 23	60 − 46

Problem Solving

Answer each question.

1. This morning the temperature was 75 degrees. This afternoon it was 83 degrees. How many degrees did it go up?

Are you to add or subtract? ____________

How many degrees did the temperature go up? ____________

1.

2. There were 45 people at a meeting. After 28 of them left, how many people were still at the meeting?

Are you to add or subtract? ____________

How many people were still at the meeting? ____________

2.

3. Renée drove 67 miles in the morning and 24 miles in the afternoon. How far did she drive?

Are you to add or subtract? ____________

How far did she drive? ____________

3.

4. Christopher is 54 inches tall. His sister is 36 inches tall. How much taller is Christopher?

Are you to add or subtract? ____________

How much taller is Christopher than his sister? ____________

4.

5. A clown has 26 orange balloons and 28 blue balloons. How many balloons is that?

Are you to add or subtract? ____________

How many orange and blue balloons are there? ____________

5.

NAME ______________________

CHAPTER 3 TEST

Add. Check each answer.

	a	*b*	*c*	*d*	*e*	*f*
1.	$36 + 7$	$45 + 9$	$8 + 23$	$17 + 7$	$8 + 44$	$58 + 6$
2.	$17 + 25$	$26 + 48$	$43 + 38$	$74 + 19$	$78 + 18$	$65 + 16$

Subtract. Check each answer.

3.	$26 - 8$	$54 - 9$	$61 - 3$	$27 - 9$	$54 - 6$	$66 - 9$
4.	$36 - 17$	$72 - 44$	$38 - 19$	$74 - 26$	$93 - 89$	$82 - 57$

Solve.

5. Fifty-four girls and 27 boys came to the meeting. How many boys and girls came to the meeting?

__________ boys and girls came to the meeting.

NAME ____________________

PRE-TEST—Addition and Subtraction

Add.

	a	*b*	*c*	*d*	*e*	*f*
1.	5 +6	50 +60	7 +8	70 +80	90 +80	70 +70
2.	53 +95	44 +74	82 +96	67 +70	55 +52	73 +86
3.	63 +78	82 +89	97 +27	56 +75	88 +88	97 +44
4.	26 +53	66 +25	74 +65	39 +87	82 +17	76 +72
5.	59 +59	73 +15	83 +67	54 +72	63 +70	35 +45

Subtract.

	a	*b*	*c*	*d*	*e*	*f*
6.	16 −7	160 −70	15 −9	150 −90	140 −60	170 −80
7.	136 −53	165 −74	154 −90	186 −93	179 −82	147 −67
8.	146 −97	158 −69	172 −85	163 −77	125 −58	116 −39
9.	176 −53	184 −35	154 −72	153 −74	146 −32	107 −40

NAME ____________________

Lesson 1 Addition and Subtraction

8 +6	8 +6 14	80 +60	80 +60 140	14 −6	14 −6 8	140 −60	140 −60 80

If 8 + 6 = 14, then 80 + 60 = ________.

If 14 − 6 = 8, then 140 − 60 = ________.

Add.

	a	*b*	*c*	*d*	*e*	*f*
1.	7 +8	70 +80	6 +9	60 +90	3 +8	30 +80
2.	7 +5	70 +50	8 +9	80 +90	4 +6	40 +60
3.	70 +40	50 +90	30 +90	70 +70	90 +40	80 +40
4.	20 +90	60 +60	70 +60	90 +10	70 +90	80 +80

Subtract.

5.	13 −5	130 −50	17 −8	170 −80	12 −6	120 −60
6.	15 −6	150 −60	14 −5	140 −50	18 −9	180 −90
7.	140 −80	110 −70	160 −80	130 −60	170 −90	120 −50
8.	130 −90	160 −70	150 −80	120 −80	140 −90	110 −40

Problem Solving

Answer each question.

1. Nicholas is on a trip of 170 kilometers. So far he has gone 90 kilometers. How many more kilometers must he go?

Are you to add or subtract? ______________________

How many more kilometers must he go? ______________

2. A school has 20 men teachers. It has 30 women teachers. How many teachers are in the school?

Are you to add or subtract? ______________________

How many teachers are in the school? __________

3. Logan weighs 70 pounds. His older brother weighs 130 pounds. How many more pounds does his older brother weigh?

Are you to add or subtract? ______________________

How many more pounds does his older brother weigh? ______________

4. Jessica has 110 pennies. Emily has 90 pennies. Jessica has how many more pennies than Emily?

Jessica has __________ more pennies than Emily.

5. Mallory sold 50 pennants on Monday and 70 on Tuesday. How many pennants did she sell in all?

Mallory sold ______________ pennants in all.

6. A bag contains 150 red and green marbles. Ninety of them are red. How many marbles are green?

______________ marbles are green.

1.	
2.	
3.	4.
5.	6.

NAME ____________________

Lesson 2 Addition

Add the ones.

$$\begin{array}{r} 43 \\ +86 \\ \hline \end{array} \qquad \begin{array}{r} 43 \\ +86 \\ \hline 9 \end{array}$$

$3 + 6 = 9$

Add the tens.

$$\begin{array}{r} 43 \\ +86 \\ \hline 129 \end{array}$$

$40 + 80 = 120$ or $100 + 20$

Add.

	a	*b*	*c*	*d*	*e*	*f*
1.	74 +62	56 +93	49 +60	57 +72	83 +35	94 +24
2.	62 +53	76 +72	34 +95	83 +43	96 +61	72 +41
3.	92 +30	74 +82	93 +92	86 +21	55 +60	34 +82
4.	65 +42	54 +82	83 +93	46 +90	93 +93	62 +64
5.	81 +58	65 +91	42 +84	35 +72	90 +70	80 +85
6.	93 +84	22 +97	45 +72	54 +54	43 +82	61 +81
7.	56 +82	62 +43	70 +76	54 +73	94 +94	85 +92

Problem Solving

Solve each problem.

1. Austin sold 96 tickets. Carmen sold 81. How many tickets did they both sell?

   Austin sold _________ tickets.

   Carmen sold _________ tickets.

   They sold a total of _________ tickets.

2. Fifty-three people live in the first building. Eighty-five people live in the second building. How many people live in both buildings?

   _________ people live in the first building.

   _________ people live in the second building.

   _________ people live in both buildings.

3. A train went 83 kilometers the first hour. The second hour it went 84 kilometers. How far did it go in the two hours?

   The first hour the train went _________ kilometers.

   The second hour it went _________ kilometers.

   In the two hours it went _________ kilometers.

4. Ninety-two train seats are filled. There are 47 empty train seats. How many train seats are there?

   There are _________ train seats.

5. Kara collected 72 stamps. Jan collected 76 stamps. How many stamps did they collect in all?

   They collected _________ stamps.

1.

2.

3.

4.

5.

NAME ______________________

Lesson 3 Subtraction

	Subtract the ones.	To subtract the tens, rename 1 hundred and 3 tens as "13 tens."	Subtract the tens.
136 −72	136 −72 4	13 ~~13~~6 −72 4	13 ~~13~~6 −72 64

Subtract.

	a	*b*	*c*	*d*	*e*	*f*
1.	147 −64	108 −72	156 −83	129 −44	175 −81	114 −42
2.	136 −86	153 −62	118 −91	124 −82	136 −43	107 −45
3.	148 −82	164 −83	186 −93	115 −72	104 −91	146 −52
4.	107 −23	139 −72	124 −30	155 −95	166 −72	124 −61
5.	118 −27	126 −55	174 −93	149 −72	108 −61	136 −94
6.	145 −92	129 −73	152 −72	164 −90	135 −62	113 −61
7.	126 −91	185 −94	137 −65	158 −86	149 −99	176 −83

Problem Solving

Solve each problem.

1. Rob had 128 centimeters of string. He used 73 centimeters of it. How much string was left?

The string was ________ centimeters long.

Rob used ________ centimeters of the string.

There were ________ centimeters of string left.

1.

2. Abby and Leigh got on a scale. The reading was "145 pounds." Leigh got off, and the reading was "75 pounds." How much does Leigh weigh?

Together they weighed ________ pounds.

Abby weighs ________ pounds.

Leigh weighs ________ pounds.

2.

3. There are 167 students in Tony's grade at school. Seventy-one of the students are girls. How many of the students are boys?

There are ________ students in all.

There are ________ girls.

There are ________ boys.

3.

4. Brittnee counted 156 sheets of paper in the package. Then she used 91 sheets. How many sheets of paper did she have left?

There were ________ sheets of paper left.

5. A jet plane has 184 passenger seats. There are 93 passengers on the plane. How many empty passenger seats are there?

There are ________ empty passenger seats.

4.

5.

NAME ______________________

Lesson 4 Addition and Subtraction

To check 75 + 61 = 136, subtract ________ from 136.

```
  75 ←
 +61
 136     These should
 −61     be the same.
  75 ←
```

To check 157 − 83 = 74, add 83 to ________.

```
 157 ←
 −83
  74     These should
 +83     be the same.
 157 ←
```

Add. Check each answer.

	a	*b*	*c*	*d*	*e*	*f*
1.	74 + 53	85 + 42	96 + 60	43 + 71	61 + 45	32 + 82
2.	91 + 82	53 + 63	96 + 51	45 + 82	32 + 96	53 + 51

Subtract. Check each answer.

3.	175 − 83	156 − 64	162 − 91	189 − 95	144 − 60	128 − 71
4.	136 − 62	165 − 83	157 − 76	128 − 61	147 − 52	104 − 21

Problem Solving

Solve each problem.

1. Derrick worked at the computer for 80 minutes in the morning. That afternoon he worked at it for 40 minutes. How many minutes did he work on the computer that day?

Are you to add or subtract? ______________

How many minutes did he work on the computer that day? ______________

1.

2. Derrick wrote a computer program that has 129 lines. He has put 91 lines in the computer so far. How many more lines does he have to put in the computer?

Are you to add or subtract? ______________

How many more lines does he have to put in the computer? ______________

2.

3. Derrick's mother uses the computer for work. Last month she used it 71 hours. This month she used it for 82 hours. How many hours did she use the computer in the last two months?

Are you to add or subtract? ______________

How many hours did she use the computer in the last two months? ______________

3.

NAME ____________________

Lesson 2 Addition

Add the ones.

```
  58        1
 +76        58
 ---       +76
           ---
             4
```

8 + 6 = 14 or 10 + 4

Add the tens.

```
  1
  58
 +76
 ---
 134
```

10 + 50 + 70 = 130 or 100 + 30

Add.

	a	b	c	d	e	f
1.	94 + 68	77 + 46	59 + 75	72 + 38	43 + 99	66 + 85
2.	87 + 85	39 + 92	66 + 46	47 + 78	75 + 55	89 + 96
3.	97 + 59	89 + 59	16 + 95	34 + 88	63 + 98	99 + 48
4.	37 + 73	94 + 28	99 + 32	58 + 95	67 + 75	29 + 85
5.	48 + 86	69 + 57	94 + 97	72 + 88	89 + 64	87 + 26
6.	54 + 88	76 + 76	89 + 98	43 + 68	96 + 29	78 + 68

Problem Solving

Solve each problem.

1. A library loaned 74 books on Monday. It loaned 87 books on Tuesday. How many books did it loan on both days?

The library loaned __________ books on Monday.

The library loaned __________ books on Tuesday.

The library loaned __________ books both days.

1.

2. Barbara read 49 pages in the morning. She read 57 pages in the afternoon. How many pages did she read in all?

Barbara read __________ pages in the morning.

Barbara read __________ pages in the afternoon.

Barbara read __________ pages in all.

2.

3. The gym is 48 feet longer than the basketball court. The basketball court is 84 feet long. How long is the gym?

The basketball court is __________ feet long.

The gym is __________ feet longer than the basketball court.

The gym is __________ feet long.

3.

4. At the circus, 84 adult tickets and 96 children's tickets were sold. How many tickets were sold?

__________ tickets were sold.

5. The team scored 66 points in the first half. They scored 68 points in the second half. How many points did they score in the game?

They scored __________ points in the game.

4.

5.

NAME ____________________

Lesson 6 Subtraction

	Rename 1 hundred and 6 ones as "10 tens and 6 ones."	Rename 10 tens and 6 ones as "9 tens and 16 ones."	Subtract the ones.	Subtract the tens.
106 −49	10 ~~10~~6 −49	9 16 ~~10~~ ~~6~~ ~~10~~6 −49	9 16 ~~10~~6 −49 7	9 16 ~~10~~6 −49 57

Subtract.

	a	b	c	d	e	f
1.	135 −86	108 −19	113 −27	125 −48	142 −59	156 −88
2.	115 −78	122 −78	171 −99	140 −55	107 −18	132 −65
3.	186 −99	153 −65	132 −93	148 −79	115 −57	142 −64
4.	153 −95	104 −37	136 −48	150 −77	162 −95	174 −86
5.	143 −85	154 −96	163 −87	132 −75	120 −61	147 −78
6.	163 −99	174 −87	126 −58	142 −95	133 −58	114 −28
7.	102 −23	175 −97	166 −97	148 −59	133 −74	121 −98

Problem Solving

Solve each problem.

1. Ms. Davis needs 180 meters of fence. She has 95 meters of fence. How many more meters of fence does she need?

Ms. Davis needs ________ meters of fence.

She has ________ meters of fence.

She needs ________ more meters of fence.

2. Aaron knows the names of 128 students at school. If 79 are girls, how many are boys?

Aaron knows the names of ________ students.

________ are girls.

________ are boys.

3. Margo's family is on a 162-kilometer trip. They have already gone 84 kilometers. How much farther do they have to go?

The trip is ________ kilometers long.

They have gone ________ kilometers.

They have ________ more kilometers to go.

4. Ian's birthday is the 29th day of the year. Karen's birthday is the 126th day. Karen's birthday is how many days after Ian's birthday?

It is ________ days after Ian's birthday.

5. Mr. Darter got 131 trading stamps at two stores. He got 84 stamps at one store. How many did he get at the other store?

Mr. Darter got ________ stamps.

1.

2.

3.

4.

5.

NAME ______________________

Lesson 7 Addition and Subtraction

Add. Check each answer.

	a	*b*	*c*	*d*	*e*	*f*
1.	54 +38	71 +56	57 +86	95 +24	42 +37	58 +26
2.	72 +96	58 +74	92 +37	48 +22	35 +43	55 +55

Subtract. Check each answer.

3.	125 −92	174 −33	165 −87	150 −90	146 −76	132 −84
4.	112 −47	118 −33	157 −26	160 −45	175 −76	153 −83
5.	198 −39	155 −97	163 −84	131 −71	111 −24	108 −39

Problem Solving

Solve each problem.

1. There are 166 people living in my apartment building. If 98 are children, how many are adults?

   There are ________ people in the building.

   There are ________ children.

   There are ________ adults.

2. There were 115 cases on a truck. The driver left 27 cases at the first stop. How many cases are still on the truck?

   ________ cases were on a truck.

   ________ cases were left at the first stop.

   ________ cases are still on the truck.

3. The bus has 84 passenger seats. All the seats are filled and there are 39 passengers standing. How many passengers are on the bus?

   The bus has ________ seats.

   There are ________ passengers standing.

   There are ________ passengers on the bus.

4. Breanne counted 63 houses on one side of the street. She counted 89 on the other side. How many houses are on the street?

   There are ________ houses on the street.

5. Lindsay had 112 balloons. She gave some of them away. She had 35 balloons left. How many balloons did she give away?

   She gave away ________ balloons.

1.
2.
3.
4.
5.

NAME ______________________

CHAPTER 4 TEST

Add or subtract. Check each answer.

	a	*b*	*c*	*d*	*e*
1.	60 +80	70 +90	85 +63	72 +54	60 +65
2.	84 +57	63 +77	82 +99	78 +78	44 +79
3.	170 −80	160 −80	153 −71	127 −82	175 −91
4.	127 −59	143 −65	166 −89	183 −95	122 −57
5.	147 −36	56 +37	175 −85	57 +89	197 −73

4

PRE-TEST—Addition and Subtraction

Add.

	a	*b*	*c*	*d*	*e*	*f*
1.	3 4 +7	8 6 +9	9 5 +7	5 6 8 +3	4 9 2 +6	3 7 5 +9
2.	10 30 40 +50	20 30 40 +60	20 40 60 +70	40 30 80 +40	50 50 20 +60	20 60 20 +40
3.	52 41 +30	26 30 +92	33 44 +57	38 46 +69	49 65 +77	27 34 +46
4.	23 23 31 +22	28 17 23 +44	91 22 34 +51	72 54 36 +21	78 52 43 +45	33 25 36 +21
5.	423 101 +324	526 345 +116	123 541 +162	752 348 +150	429 316 541 +302	324 115 462 +115

Subtract.

	a	*b*	*c*	*d*	*e*
6.	752 −341	673 −424	583 −193	765 −489	605 −329
7.	4723 −221	5806 −447	3924 −163	7811 −912	6425 −587

NAME ____________________

Lesson 1 Addition

	Add the ones.			Add the tens.
67 98 +83	7 8 → 15 +3	15 +3 18 or 10 + 8	1 67 98 + 83 8	1 67 98 + 83 248

Add.

	a	b	c	d	e	f
1.	4 5 + 7	6 8 + 9	5 2 + 8	9 8 + 3	4 6 + 5	7 7 + 6
2.	10 40 30 + 50	20 60 50 + 60	10 20 90 + 40	20 40 30 + 70	10 50 60 + 40	20 70 50 + 80
3.	44 35 + 57	66 58 + 59	25 92 + 48	49 38 + 73	54 66 + 45	77 57 + 86
4.	25 32 + 41	27 35 + 42	55 55 + 55	32 44 + 28	75 16 + 58	22 14 + 91
5.	57 28 + 36	42 54 + 78	79 34 + 29	68 78 + 88	25 36 + 42	53 26 + 13
6.	45 18 + 52	61 29 + 58	83 76 + 19	49 42 + 43	37 67 + 26	98 16 + 35

Problem Solving

NATIONAL LEAGUE TEAM STANDINGS		
TEAM	WON	LOST
CUBS	72	43
CARDINALS	69	48
METS	64	52
PIRATES	58	55
PHILLIES	44	68
EXPOS	37	79

Solve each problem.

1. How many games have been won by the first three teams in the National League?

   The Cubs have won ______________ games.

   The Cardinals have won ______________ games.

   The Mets have won ______________ games.

   Together they have won ______________ games.

2. How many games have been lost by the last three teams in the National League?

   The Pirates have lost ______________ games.

   The Phillies have lost ______________ games.

   The Expos have lost ______________ games.

   Together they have lost ______________ games.

3. How many games have been won by the Cubs, Mets, Phillies, and Expos?

   They have won ______________ games.

4. How many games have the Cubs, Cardinals, and Pirates lost?

   They have lost ________________________ games.

1.

2.

3.

4.

NAME ______________________

Lesson 2 Addition

	Add the ones.	Add the tens.	Add the hundreds.
	1	2 1	2 1
642	642	642	642
156	156	156	156
275	275	275	275
+143	+143	+143	+143
	6	16	1216

2 + 6 + 5 + 3 = ___ | 10 + 40 + 50 + 70 + 40 = ____ | 200 + 600 + 100 + 200 + 100 = ____

16 = 10 + _____ | 210 = 200 + _______ | 1200 = 1000 + ________

Add.

	a	*b*	*c*	*d*	*e*	*f*
1.	372 456 +174	382 154 +283	231 336 +136	152 443 +178	321 305 +238	143 116 +212
2.	425 641 +703	443 217 +602	613 247 +138	574 142 +281	382 425 +678	392 456 +731
3.	728 365 +428	639 752 +417	618 304 +120	856 174 +372	564 345 +654	224 305 +406
4.	421 145 162 +231	178 214 103 +407	513 223 641 +412	421 146 273 +154	762 531 444 +258	372 541 635 +413
5.	603 254 316 +222	425 245 542 +254	631 211 431 +222	731 240 635 +214	245 361 524 +113	284 563 711 +245

Problem Solving

Solve each problem.

1. The local theater had a special Saturday movie. They sold 175 tickets to men, 142 to women, and 327 to children. How many tickets did they sell in all?

They sold ____________________ tickets to men.

They sold ____________________ tickets to women.

They sold ____________________ tickets to children.

They sold ____________________ tickets in all.

1.

2. In the local high school there are 768 boys, 829 girls, and 107 teachers. How many teachers and students are there in all?

There are ____________________ boys.

There are ____________________ girls.

There are ____________________ teachers.

There are ____________________ teachers and students in all.

2.

3. The number of people living in 4 different apartment buildings is 203, 245, 268, and 275. How many people live in all 4 buildings?

__________ people live in all 4 buildings.

4. A living room floor has 195 tiles. A bedroom floor has 168 tiles. A kitchen floor has 144 tiles. How many tiles are in the 3 rooms?

There are __________ tiles in these 3 rooms.

3.

4.

NAME ______________________

Lesson 3 Subtraction

	Rename 40 as "3 tens and 10 ones." Then subtract the ones.	Rename 7 hundreds and 3 tens as "6 hundreds and 13 tens." Then subtract the tens.	Subtract the hundreds.
740 −271	7 ~~4~~(3) ~~0~~(10) −271 9	~~7~~(6) ~~4~~(~~3~~ 13) ~~0~~(10) −271 69	~~7~~(6) ~~4~~(~~3~~ 13) ~~0~~(10) −271 469

Subtract.

	a	*b*	*c*	*d*	*e*	*f*
1.	534 −273	263 −154	758 −439	450 −261	536 −347	274 −154
2.	463 −372	782 −234	594 −287	681 −382	384 −175	806 −764
3.	764 −137	635 −447	492 −113	780 −152	444 −235	562 −357
4.	836 −257	944 −256	758 −167	504 −235	672 −285	892 −284
5.	945 −463	378 −126	564 −243	839 −257	245 −146	776 −382
6.	805 −308	900 −750	764 −345	840 −426	955 −765	436 −327

Problem Solving

Solve each problem.

1. Babe Ruth hit 714 home runs. Henry (Hank) Aaron hit 755 home runs. How many more home runs did Hank Aaron hit than Babe Ruth?

Babe Ruth hit ______________ home runs.

Hank Aaron hit ______________ home runs.

Hank Aaron hit ______________ more home runs than Babe Ruth.

2. A train has 850 seats. There are 317 empty seats. How many people are seated?

The train has __________ seats.

__________ seats are empty.

There are __________ people seated.

3. Hoover Dam is 726 feet high. Folsom Dam is 340 feet high. How much higher is Hoover Dam than Folsom Dam?

Hoover Dam is __________ feet high.

Folsom Dam is __________ feet high.

Hoover Dam is __________ feet higher than Folsom Dam.

4. The quarterback threw 247 passes. Only 138 passes were caught. How many were not caught?

__________ passes were not caught.

5. A meeting room can hold 443 people. There are 268 people in the room now. How many more people can it hold?

It can hold __________ more people.

1.

2.

3.

4.

5.

NAME ____________________

Lesson 4 Subtraction

	Subtract the ones.	Rename 2 hundreds and 5 tens as "1 hundred and 15 tens." Subtract the tens.	Rename 4 thousands and 1 hundred as "3 thousands and 11 hundreds." Subtract the hundreds.	Subtract the thousands.
4253 − 281	4253 − 281 = 2	4253 (1 15) − 281 = 72	4253 (3 11 15) − 281 = 972	4253 (3 11 15) − 281 = 3972

Subtract.

	a	*b*	*c*	*d*	*e*
1.	7543 − 211	6813 − 402	7254 − 132	4936 − 726	2815 − 813
2.	3562 − 235	4253 − 147	6541 − 538	3473 − 255	5496 − 339
3.	3710 − 340	9642 − 271	3817 − 454	5216 − 182	3847 − 377
4.	4295 − 724	4007 − 805	8281 − 470	5554 − 644	6382 − 882
5.	5986 − 537	2413 − 829	4507 − 758	3154 − 205	2604 − 834
6.	8329 − 475	7604 − 829	3987 − 988	4205 − 736	1383 − 529

Problem Solving

Solve each problem.

1. Ms. Ramos bought a car that cost 3,165 dollars. She paid 875 dollars. How much does she still owe?

   The new car cost ______________ dollars.

   Ms. Ramos paid ______________ dollars.

   She still owes ______________ dollars.

2. Mount Whitney is 4418 meters high. Mount Davis is 979 meters high. How much higher is Mount Whitney?

   Mount Whitney is __________ meters high.

   Mount Davis is __________ meters high.

   Mount Whitney is __________ meters higher.

3. There are 1,156 students enrolled in a school. Today 219 students are absent. How many are present?

   __________ students are present.

4. There are 5,280 feet in a mile. John walked 895 feet. How many more feet must he go to walk a mile?

   He must go __________ more feet to walk a mile.

5. Albertito's family went 2198 kilometers in 5 days. They went 843 kilometers the first 2 days. How many kilometers did they go in the last 3 days?

   They went __________ kilometers in the last three days.

6. There are 1,255 people on a police force. If 596 are women, how many are men?

   There are __________ men.

1.	
2.	
3.	4.
5.	6.

NAME ______________________

CHAPTER 5 TEST

Add.

	a	b	c	d	e
1.	3 2 +5	4 7 +6	8 4 +9	70 30 50 +40	50 70 80 +30
2.	35 24 +20	57 13 +28	74 82 +36	23 32 58 +42	42 53 64 +70
3.	421 312 +148	623 174 +162	473 126 +248	326 112 224 +607	526 381 426 +543

Subtract.

	a	b	c	d
4.	765 −243	290 −183	846 −354	846 −297
5.	5836 −314	7542 −275	6039 −268	2560 −764

Solve each problem.

6. Four girls earned the following points in a contest: 145, 387, 245, and 197. What was the total number of points earned?

The total number of points was ______________.

7. Jay's new car was driven 837 miles. Tara's new car was driven 3,275 miles. How many more miles was Tara's new car driven than Jay's?

Tara's car was driven ______________ more miles.

6.

7.

NAME ____________________

PRE-TEST—Measurement

Complete the following.

	a	*b*
1.	There are ________ days in a year.	4:10 means 10 minutes after ________.
2.	There are ________ days in a leap year.	3:50 means 10 minutes to ________.
3.	There are ________ days in April.	5:45 means ________ minutes after 5.
4.	There are ________ days in March.	5:45 means ________ minutes to 6.

Complete the following as shown.

	a	*b*	*c*
5.	XI = 11	V = ________	IV = ________
6.	XVII = ________	XXVI = ________	XIX = ________
7.	7 = VII	10 = ________	9 = ________
8.	24 = ________	31 = ________	25 = ________

Add or subtract.

	a	*b*	*c*	*d*	*e*
9.	\$5.20 + 6.89	\$12.65 + 1.25	46¢ + 37¢	29¢ 37¢ + 28¢	\$14.50 0.28 + 3.73
10.	\$16.50 − 3.25	\$14.75 − 2.90	\$7.40 − 0.84	56¢ − 38¢	97¢ − 50¢

Solve.

11. Ms. Romanez bought a saw for \$21.95 and a hammer for \$9.49. She paid \$1.88 tax. How much was her total bill?

Her total bill was ____________________.

NAME ____________________

Lesson 1 Reading Our Calendar

January

S	M	T	W	T	F	S
						1
2	3	4	5	6	7	8
9	10	11	12	13	14	15
16	17	18	19	20	21	22
23	24	25	26	27	28	29
30	31					

Febuary

S	M	T	W	T	F	S
		1	2	3	4	5
6	7	8	9	10	11	12
13	14	15	16	17	18	19
20	21	22	23	24	25	26
27	28					

March

S	M	T	W	T	F	S
		1	2	3	4	5
6	7	8	9	10	11	12
13	14	15	16	17	18	19
20	21	22	23	24	25	26
27	28	29	30	31		

April

S	M	T	W	T	F	S
					1	2
3	4	5	6	7	8	9
10	11	12	13	14	15	16
17	18	19	20	21	22	23
24	25	26	27	28	29	30

May

S	M	T	W	T	F	S
1	2	3	4	5	6	7
8	9	10	11	12	13	14
15	16	17	18	19	20	21
22	23	24	25	26	27	28
29	30	31				

June

S	M	T	W	T	F	S
			1	2	3	4
5	6	7	8	9	10	11
12	13	14	15	16	17	18
19	20	21	22	23	24	25
26	27	28	29	30		

July

S	M	T	W	T	F	S
					1	2
3	4	5	6	7	8	9
10	11	12	13	14	15	16
17	18	19	20	21	22	23
24	25	26	27	28	29	30
31						

August

S	M	T	W	T	F	S
	1	2	3	4	5	6
7	8	9	10	11	12	13
14	15	16	17	18	19	20
21	22	23	24	25	26	27
28	29	30	31			

September

S	M	T	W	T	F	S
				1	2	3
4	5	6	7	8	9	10
11	12	13	14	15	16	17
18	19	20	21	22	23	24
25	26	27	28	29	30	

October

S	M	T	W	T	F	S
						1
2	3	4	5	6	7	8
9	10	11	12	13	14	15
16	17	18	19	20	21	22
23	24	25	26	27	28	29
30	31					

November

S	M	T	W	T	F	S
		1	2	3	4	5
6	7	8	9	10	11	12
13	14	15	16	17	18	19
20	21	22	23	24	25	26
27	28	29	30			

December

S	M	T	W	T	F	S
				1	2	3
4	5	6	7	8	9	10
11	12	13	14	15	16	17
18	19	20	21	22	23	24
25	26	27	28	29	30	31

There are 365 days in the calendar year shown. Every four years, there are 366 days in a year. It is called a **leap year.** Only in a leap year is there a February 29.

There are __31__ days in March. There are ______ days in June.

March 1 is on __Tuesday__. June 1 is on ______________.

On the calendar above, April has __4__ Sundays and ____ Saturdays.

Answer each question. Use the calendar to help you.

	a	*b*
1.	How many days are in July? ______________	On what day is July 1? ______________
2.	How many Tuesdays are in November? ______________	How many Wednesdays are in November? ______________
3.	How many months have 30 days? ______________	How many months have 31 days? ______________
4.	What date is the 3rd Thursday in August? ______________	What date is the 2nd Monday in April? ______________
5.	How many days of the year have passed when we reach May 1? ______________	What date falls forty-five days before December 25? ______________

NAME ______________________________

Lesson 2 Telling Time

7:10

7:10 is read "seven ten" and means "10 minutes after 7."

3:40

3:40 is read "three forty" and means "40 minutes after 3" or "20 minutes to 4."

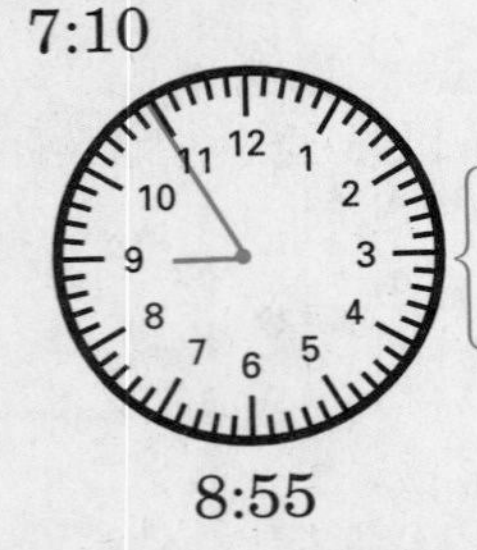

8:55

8:55 is read "eight fifty-five" and means "55 minutes after ______"

or "______ minutes to ______."

Complete the following.

	a	*b*
1.	3:05 means ______ minutes after ______.	6:50 means ______ minutes to ______.
2.	10:20 means ______ minutes after ______.	11:35 means ______ minutes to ______.
3.	8:45 means ______ minutes after ______.	8:45 means ______ minutes to ______.
4.	5:30 means ______ minutes after ______.	5:30 means ______ minutes to ______.
5.	1:10 means ______ minutes after ______.	12:55 means ______ minutes to ______.

For each clockface, write the numerals that name the time.

6.

a

________:________

b

________:________

c

________:________

d

________:________

7.

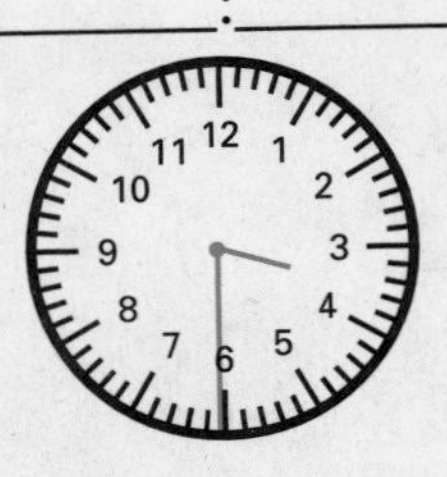

________:________

________:________

________:________

________:________

NAME ______________________

Lesson 3 Roman Numerals

I means 1. V means 5. X means 10.

II means 1 + 1 or 2.

VI means 5 + 1 or 6.

XXV means 10 + 10 + 5 or 25.

III means 1 + 1 + 1 or 3.

IV means 5 − 1 or 4.

IX means 10 − 1 or 9.

VII means 5 + 1 + _____ or _____.

XXI means 10 + _____ + 1 or _____.

XIV means _____ + 4 or _____.

XIX means _____ + 9 or _____.

Complete the following as shown.

	a	*b*	*c*	*d*
1.	XXIV = 24	XX = _____	XII = _____	VIII = _____
2.	IV = _____	XXVI = _____	XVII = _____	XXXI = _____
3.	XXXVI = _____	XXIX = _____	XI = _____	XXXIII = _____
4.	XVIII = _____	IX = _____	XXXIV = _____	XIII = _____
5.	V = _____	XXV = _____	VI = _____	XXI = _____
6.	XXXVIII = _____	XXXV = _____	XXVII = _____	XVI = _____
7.	XXIII = _____	XXXVII = _____	XIV = _____	XXXII = _____

Write a Roman numeral for each of the following.

	a	*b*	*c*
8.	3 = _____	7 = _____	15 = _____
9.	19 = _____	22 = _____	28 = _____
10.	30 = _____	20 = _____	39 = _____

NAME ____________________

Lesson 4 Money

1 penny	1 nickel	1 dime	1 quarter	1 dollar
1 cent	5 cents	10 cents	25 cents	100 cents
1¢ or $0.01	5¢ or $0.05	10¢ or $0.10	25¢ or $0.25	$1.00

25 pennies have a value of ___25___ cents or ___1___ quarter.

5 pennies have a value of ________ cents or ________ nickel.

$2.57 means ___2___ dollars and ___57___ cents.

$3.45 means ________ dollars and ________ cents.

Complete the following.

1. 10 pennies have a value of ________ cents or ________ nickels.
2. 10 pennies have a value of ________ cents or ________ dime.
3. 20 pennies have a value of ________ cents or ________ dimes.
4. 15 pennies have a value of ________ cents or ________ nickels.
5. 20 pennies have a value of ________ cents or ________ nickels.

Complete the following as shown.

6. $14.05 means ___14___ dollars and ___5___ cents.
7. $12.70 means ________ dollars and ________ cents.
8. $8.14 means ________ dollars and ________ cents.
9. $0.65 means ________ dollars and ________ cents.
10. $10.01 means ________ dollars and ________ cent.

NAME ______________________

Lesson 5 Money

	$12.00				
$9.05	0.45	45¢	$0.75	$14.08	$13.00
+6.98	+3.16	+38¢	+0.38	−7.25	−6.05
$16.03	$15.61	83¢	$1.13	$6.83	$6.95

Add or subtract as usual.

Put a decimal point (.) and a $ or ¢ in the answer.

Be sure to line up the decimal points.

Add or subtract.

	a	*b*	*c*	*d*	*e*
1.	$0.36 +12.40	$3.75 +1.46	$1.36 +40.00	37¢ +58¢	$4.35 +0.27
2.	$5.20 −3.18	$12.64 −5.08	$3.00 −0.54	88¢ −76¢	$24.42 −1.68
3.	$4.23 16.90 +0.89	$7.25 0.40 +4.42	$8.05 12.16 +0.58	47¢ 18¢ +25¢	$0.08 3.67 +14.30
4.	$15.40 −3.62	$5.70 −2.08	$11.30 −0.86	91¢ −75¢	$17.20 −4.06
5.	$27.00 −13.45	$65.21 +3.80	$0.12 +1.88	47¢ −19¢	$3.00 −1.78
6.	$16.49 +28.98	$40.60 −7.56	$5.00 −2.72	38¢ +35¢	$8.75 +0.64

Problem Solving

Solve each problem.

1. Caitlin's mother bought a dress for \$22.98 and a blouse for \$17.64. How much did these items cost altogether?

They cost __________ altogether.

2. Find the total cost of a basketball at \$18.69, a baseball at \$8.05, and a football at \$24.98.

The total cost is __________.

3. Jeremy has \$2.50. Landon has \$1.75. Jeremy has how much more money than Landon?

Jeremy has __________ more than Landon.

4. In problem **2,** how much more does the basketball cost than the baseball? How much more does the football cost than the basketball?

The basketball costs __________ more than the baseball.

The football costs __________ more than the basketball.

5. Alexandra saved \$4.20 one week, \$0.90 the next week, and \$2.05 the third week. How much money did she save during these 3 weeks?

Alexandra saved __________ in 3 weeks.

6. Mr. Lewis paid \$4.45 for fruit. He paid \$0.99 for potatoes. The tax was \$.33. How much was the total bill?

His total bill was __________.

7. Tyler wants to buy a 95¢ whistle. He now has 68¢. How much more money does he need to buy the whistle?

Tyler needs __________ more.

1.

2.

3.

4.

5.

6.

7.

NAME ______________________

CHAPTER 6 TEST

Answer each question. Use the calendar to help you.

1. How many days are in May? ____________

2. On what day is May 4? ____________

May

S	M	T	W	T	F	S
		1	2	3	4	5
6	7	8	9	10	11	12
13	14	15	16	17	18	19
20	21	22	23	24	25	26
27	28	29	30	31		

For each clockface, write the numerals that name the time.

3.

a

______ : ______

b

______ : ______

c

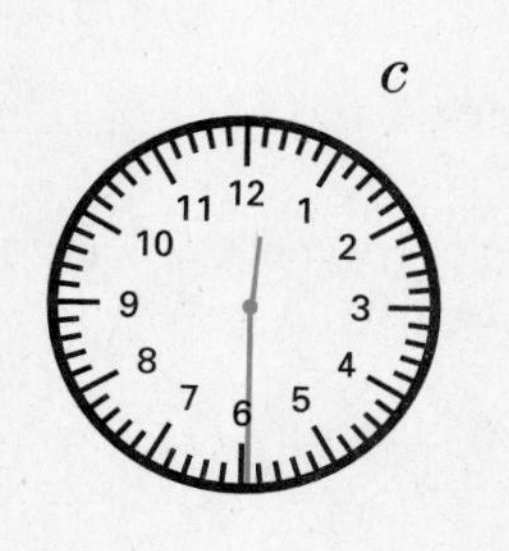

______ : ______

Complete the following as shown.

	a	*b*	*c*
4.	XVI = 16	IX = ______	XXXII = ______
5.	14 = XIV	8 = ______	29 = ______

Add or subtract.

	a	*b*	*c*	*d*	*e*
6.	$1 5.3 2 +1 6.4 5	$3.2 4 +0.7 3	4 2¢ +5 4¢	1 6¢ 3 7¢ +2 0¢	$1 3.4 0 0.6 2 +1.6 8
7.	$3.5 2 −2.1 7	$1 3.1 4 −5.3 3	9 3¢ −3 9¢	$1 7.5 0 −1.0 9	$5.1 4 −1.0 8

Solve.

8. Maria needs $54.68 to buy a coat she wants. She now has $50.75. How much more money does she need to buy the coat?

Maria needs ______________________ more.

NAME ______________________

PRE-TEST—Multiplication

Multiply.

	a	*b*	*c*	*d*	*e*	*f*
1.	5 ×2	7 ×2	2 ×2	6 ×2	4 ×2	9 ×2
2.	3 ×3	5 ×3	4 ×3	7 ×3	9 ×3	2 ×3
3.	7 ×0	5 ×0	0 ×4	0 ×6	3 ×0	0 ×8
4.	3 ×1	7 ×1	1 ×4	1 ×1	5 ×1	1 ×8
5.	7 ×4	3 ×4	9 ×4	6 ×4	5 ×4	4 ×4
6.	8 ×5	6 ×5	9 ×5	4 ×5	3 ×5	2 ×5
7.	9 ×0	8 ×4	6 ×3	0 ×1	5 ×5	0 ×3
8.	1 ×9	2 ×4	1 ×2	7 ×5	8 ×3	2 ×1
9.	3 ×2	1 ×3	0 ×7	8 ×2	1 ×6	1 ×5

NAME ____________________

Lesson 1 Multiplication

2×3 is read "two times three."
3×2 is read "three times two."
4×5 is read "four times five."

2×3 means $3 + 3$.
3×2 means $2 + 2 + 2$.
4×5 means $5 + 5 + 5 + 5$.

3×6 is read "three times six."

3×6 means ____________________

2×7 is read "two times seven."

2×7 means ____________________

Complete the following as shown.

1. 2×5 is read "two times five"

2. 3×4 is read ____________________

3. 5×2 is read ____________________

4. 4×8 is read ____________________

5. 4×7 is read ____________________

Complete the following as shown.

	a	*b*
6.	2×4 means $4 + 4$	4×2 means $2 + 2 + 2 + 2$
7.	3×5 means ________	5×3 means ________
8.	3×7 means ________	7×3 means ________
9.	4×6 means ________	6×4 means ________
10.	2×8 means ________	8×2 means ________
11.	3×9 means ________	9×3 means ________

NAME ______________________________

Lesson 2 Multiplication

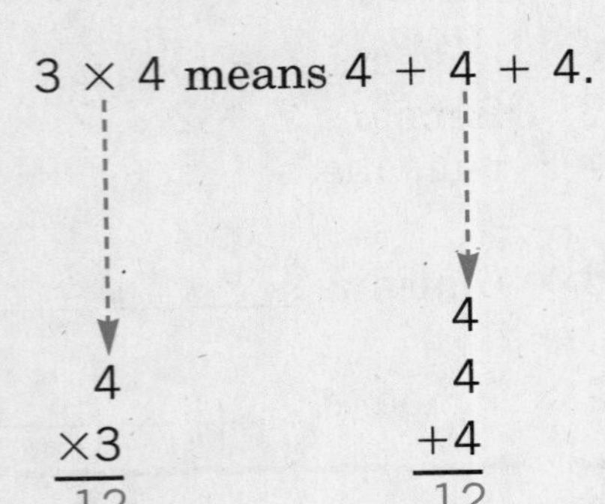

3 × 4 means 4 + 4 + 4.

4 ×3 12	4 4 +4 12

4 × 3 means 3 + 3 + 3 + 3.

3 ×4 12	3 3 3 +3 12

Add or multiply.

	a	*b*	*c*	*d*	*e*	*f*
1.	8 +8	8 ×2	4 +4	4 ×2	5 +5	5 ×2
2.	6 +6	6 ×2	7 +7	7 ×2	2 +2	2 ×2
3.	9 +9	9 ×2	3 +3	3 ×2	1 +1	1 ×2
4.	2 2 +2	2 ×3	3 3 +3	3 ×3	4 4 +4	4 ×3
5.	5 5 +5	5 ×3	6 6 +6	6 ×3	7 7 +7	7 ×3
6.	8 8 +8	8 ×3	9 9 +9	9 ×3	1 1 +1	1 ×3

NAME ______________________

Lesson 3 Multiplication

$$\begin{array}{r} 1 \\ \times 0 \\ \hline 0 \end{array} \quad \begin{array}{r} 2 \\ \times 0 \\ \hline 0 \end{array} \quad \begin{array}{r} 0 \\ \times 3 \\ \hline 0 \end{array} \quad \begin{array}{r} 0 \\ \times 4 \\ \hline 0 \end{array} \qquad \begin{array}{r} 0 \\ \times 1 \\ \hline 0 \end{array} \quad \begin{array}{r} 1 \\ \times 1 \\ \hline 1 \end{array} \quad \begin{array}{r} 2 \\ \times 1 \\ \hline 2 \end{array} \quad \begin{array}{r} 1 \\ \times 3 \\ \hline 3 \end{array}$$

Multiply.

	a	*b*	*c*	*d*	*e*	*f*
1.	0×2	9×1	1×7	6×0	1×5	0×7
2.	4×0	8×1	1×4	0×9	7×0	6×1
3.	5×0	0×8	5×1	1×6	1×1	8×0
4.	1×7	0×4	3×0	9×0	7×1	1×5
5.	0×7	1×9	1×6	0×5	1×0	2×1
6.	1×4	1×8	4×0	8×1	0×6	0×3
7.	0×9	6×1	0×2	9×1	0×1	3×1
8.	1×2	6×0	7×0	1×3	4×1	0×0

Problem Solving

Solve each problem.

1. Molly bought 2 baseball cards. Each baseball card cost 9 cents. How much did Molly pay for the baseball cards?

Molly bought __________ baseball cards.

Each baseball card cost __________ cents.

Molly paid __________ cents for the baseball cards.

2. Cody bought 2 football cards. They cost 6 cents each. How much did Cody pay for the football cards?

Cody bought __________ football cards.

One football card cost __________ cents.

Cody paid __________ cents for the football cards.

3. There are 8 cards in each pack. How many cards are in 3 packs?

__________ cards are in three packs.

4. One basketball card costs 5 cents. How much will 8 basketball cards cost?

Eight basketball cards will cost __________ cents.

1.

2.

3.

4.

NAME ____________________

Lesson 4 Multiplication

4-column

×	0	1	2	3	4	5	6	7	8	9
0	0	0	0	0	0	0	0	0	0	0
1	0	1	2	3	4	5	6	7	8	9
2	0	2	4	6	8	10	12	14	16	18
3	0	3	6	9	12	15	18	21	24	27
4	0	4	8	12	16	20	24	28	32	36
5	0	5	10	15	20	25	30	35	40	45
6	0	6	12	18	(24)	30				
7	0	7	14	21	28	35				
8	0	8	16	24	32	40				
9	0	9	18	27	36	45				

6-row

$$\begin{array}{r} 6 \\ \times 4 \\ \hline 24 \end{array}$$

6 → Find the 6 -row.
×4 → Find the 4 -column.
24 ← The product is named where the 6-row and 4-column meet.

Multiply.

	a	*b*	*c*	*d*	*e*	*f*
1.	5×4	8×4	7×5	6×5	2×4	4×3
2.	5×5	6×3	9×4	1×4	0×5	4×4
3.	3×5	7×4	2×5	4×2	8×5	9×2
4.	5×3	3×3	8×2	0×4	3×2	5×2
5.	6×4	8×3	4×1	5×0	5×1	6×2
6.	9×5	4×0	3×4	7×2	7×3	1×5

Problem Solving

Solve each problem.

1. Ashley wants to buy 5 erasers. They cost 9 cents each. How much will she have to pay?

Ashley wants to buy _________ erasers.

One eraser costs _________ cents.

Ashley will have to pay _________ cents.

2. There are 5 rows of mailboxes. There are 7 mailboxes in each row. How many mailboxes are there in all?

There are _________ mailboxes in each row.

There are _________ rows of mailboxes.

There are _________ mailboxes in all.

3. Milton, the pet monkey, eats 4 meals every day. How many meals does he eat in a week?

There are _________ days in a week.

Milton eats _________ meals every day.

Milton eats _________ meals in a week.

4. In a baseball game each team gets 3 outs per inning. How many outs does each team get in a 5-inning game?

There are _________ innings in the game.

Each team gets _________ outs per inning.

The team gets _________ outs in the 5-inning game.

5. Cameron has gained 4 pounds in each of the past 5 months. How much weight has he gained?

Cameron has gained _________ pounds in 5 months.

1.

2.

3.

4.

5.

NAME ______________________

Lesson 5 Multiplication

Multiply.

	a	*b*	*c*	*d*	*e*	*f*
1.	0×8	4×2	8×5	7×3	6×1	7×0
2.	1×1	9×2	4×4	3×5	6×5	1×4
3.	0×6	1×2	4×0	8×2	9×5	5×5
4.	8×4	6×3	1×5	9×0	2×1	7×2
5.	5×3	7×4	4×5	3×2	9×3	8×1
6.	6×0	3×1	6×2	0×0	2×3	9×4
7.	7×5	8×3	1×0	0×3	4×1	6×4
8.	5×4	2×2	9×1	1×7	2×4	3×3
9.	1×9	2×0	5×2	3×4	2×5	4×3

Problem Solving

Solve each problem.

1. Neal has 6 books. Each book weighs 1 kilogram. What is the weight of all the books?

Neal has _________ books.

Each book weighs _________ kilogram.

The six books weigh _________ kilograms.

2. A basketball game has 4 time periods. Kate's team is to play 8 games. How many periods will her team play?

Kate's team is to play _________ games.

Each game has _________ time periods.

Kate's team will play _________ time periods in all.

3. Meagan works 8 hours every day. How many hours does she work in 5 days?

She works _________ hours in 5 days.

4. Shane can ride his bicycle 5 miles in an hour. At that speed how far could he ride in 2 hours?

Shane could ride _________ miles in 2 hours.

5. Calvin bought 5 bags of balloons. Each bag had 6 balloons. How many balloons did he buy?

Calvin bought _________ balloons in all.

6. Kristen can build a model car in 3 hours. How long would it take her to build 4 model cars?

Kristen could build 4 model cars in _________ hours.

1.

2.

3.

4.

5.

6.

NAME ____________________

CHAPTER 7 TEST

Multiply.

	a	b	c	d	e
1.	1×6	7×4	9×0	3×4	9×5
2.	4×3	7×3	0×6	1×4	6×2
3.	9×2	8×5	9×3	3×2	4×4
4.	0×1	5×5	9×4	3×0	4×5

Solve each problem.

5. Nathan bought 5 boxes of pencils. There are 6 pencils in each box. How many pencils did he buy?

Nathan bought __________ boxes of pencils.

There are __________ pencils in each box.

He bought __________ pencils in all.

6. Erin is to put 4 apples in each bag. How many apples does she need to fill 8 bags?

Erin needs __________ apples in all.

7. Troy bought 3 boxes of crayons. Each box held 8 crayons. How many crayons did he buy?

Troy bought __________ crayons.

5.

6.

7.

7

PRE-TEST—Multiplication

Multiply.

	a	*b*	*c*	*d*	*e*	*f*
1.	7×6	6×6	4×6	8×6	5×6	9×6
2.	8×7	4×7	9×7	7×7	6×7	5×7
3.	9×8	5×8	7×8	8×8	6×8	4×8
4.	6×9	9×9	5×9	8×9	4×9	7×9

Solve each problem.

5. Luke set up 9 rows of chairs. He put 9 chairs in each row. How many chairs did he use?

Luke used ________ chairs.

6. Bethany's dad works 8 hours every day. How many hours would he work in 7 days?

He would work ________ hours in 7 days.

7. There are 9 players on a team. How many players are there on 7 teams?

There are ________ players in all.

8. Brent puts 6 apples into each bag. How many apples does he need to fill 7 bags?

He would need ________ apples.

5.	6.
7.	8.

NAME ______________________

Lesson 1 Multiplication

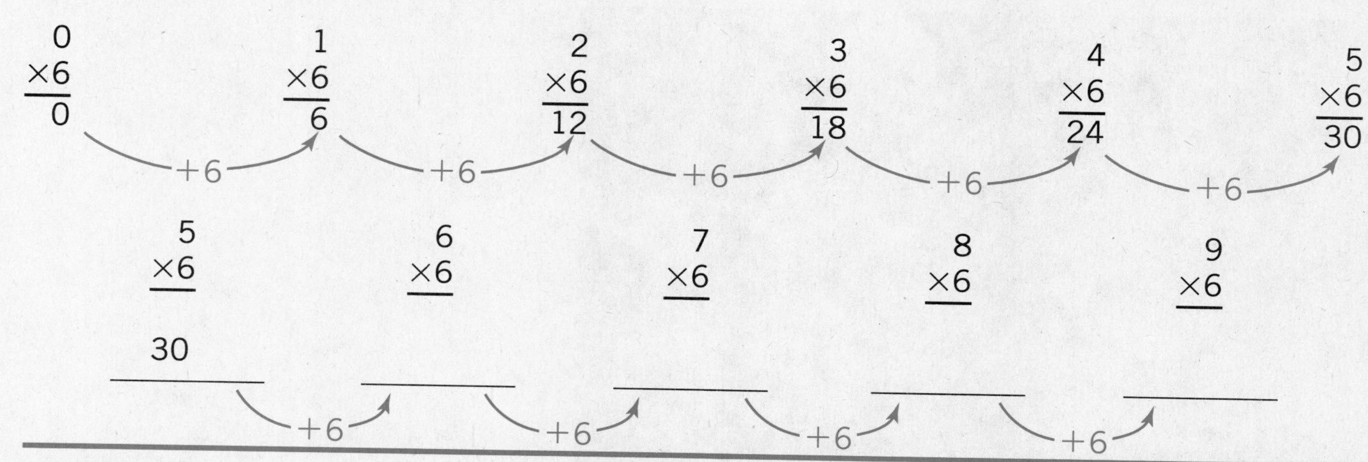

Multiply.

	a	b	c	d	e	f
1.	5 ×6	3 ×3	9 ×6	8 ×4	0 ×5	5 ×3
2.	8 ×5	2 ×6	6 ×4	5 ×5	3 ×4	6 ×1
3.	4 ×3	7 ×4	6 ×5	8 ×6	9 ×4	8 ×3
4.	0 ×3	3 ×6	1 ×5	4 ×4	2 ×3	4 ×5
5.	7 ×6	9 ×5	7 ×5	0 ×4	5 ×2	2 ×4
6.	6 ×6	6 ×2	5 ×4	1 ×3	3 ×5	0 ×6
7.	1 ×4	2 ×5	1 ×6	9 ×3	7 ×3	4 ×6

Problem Solving

Solve each problem.

1. There are 6 rows of cactus plants. Each row has 4 plants. How many cactus plants are there in all?

   There are __________ rows of cactus plants.

   There are __________ cactus plants in each row.

   There are __________ cactus plants in all.

1.

2. There are 8 marigold plants in each row. There are 6 rows. How many marigold plants are there?

   There are __________ marigold plants in each row.

   There are __________ rows of marigold plants.

   There are __________ marigold plants in all.

2.

3. There are 6 rosebushes in each row. There are 9 rows. How many rosebushes are there?

   There are __________ rosebushes in each row.

   There are __________ rows of rosebushes.

   There are __________ rosebushes in all.

3.

NAME ____________________

Lesson 2 Multiplication

4×7	5×7	6×7	7×7	8×7	9×7
28	35	42			

+7 +7 +7 +7 +7

4×8	5×8	6×8	7×8	8×8	9×8
32	40	48			

+8 +8 +8 +8 +8

Multiply.

	a	*b*	*c*	*d*	*e*	*f*
1.	7×7	7×6	6×8	3×7	9×8	0×7
2.	8×8	8×5	1×8	7×3	3×8	1×7
3.	8×0	8×6	5×8	9×7	7×1	6×7
4.	7×8	5×7	7×0	7×2	7×5	0×8
5.	8×7	2×8	8×4	7×4	8×3	4×8
6.	9×4	8×1	4×7	8×2	2×7	9×6

Problem Solving

Solve each problem.

1. In Tori's building there are 7 floors. There are 9 apartments on each floor. How many apartments are in the building?

There are __________ floors in this building.

There are __________ apartments on each floor.

There are __________ apartments in this building.

2. The science club meets 4 times each month. The club meets for 7 months. How many meetings will the science club have?

The science club meets ________ times each month.

The club meets for __________ months.

The club will have __________ meetings in all.

3. Each bag of corn weighs 8 kilograms. There are 7 bags. How much do the bags weigh in all?

Each bag weighs __________ kilograms.

There are __________ bags.

The bags weigh __________ kilograms in all.

4. There are 7 days in a week. How many days are there in 5 weeks?

There are __________ days in 5 weeks.

5. Brenda walks 6 blocks each day going to and from school. How many blocks does she walk going to and from school in 7 days?

Brenda walks __________ blocks in 7 days.

1.

2.

3.

4.

5.

NAME ________________________

Lesson 3 Multiplication

9 ⟶ Find the 9 -row.
×9 ⟶ Find the 9 -column.
81 ⟵ The product is named where the 9-row and 9-column meet.

9-column

9-row

×	0	1	2	3	4	5	6	7	8	9
0	0	0	0	0	0	0	0	0	0	0
1	0	1	2	3	4	5	6	7	8	9
2	0	2	4	6	8	10	12	14	16	18
3	0	3	6	9	12	15	18	21	24	27
4	0	4	8	12	16	20	24	28	32	36
5	0	5	10	15	20	25	30	35	40	45
6	0	6	12	18	24	30	36	42	48	54
7	0	7	14	21	28	35	42	49	56	63
8	0	8	16	24	32	40	48	56	64	72
9	0	9	18	27	36	45	54	63	72	(81)

Multiply.

	a	*b*	*c*	*d*	*e*	*f*
1.	7 ×9	9 ×6	8 ×8	3 ×7	2 ×9	7 ×3
2.	8 ×5	8 ×7	4 ×9	6 ×8	7 ×0	7 ×7
3.	9 ×9	1 ×8	0 ×9	6 ×9	7 ×8	2 ×8
4.	4 ×8	0 ×7	4 ×7	9 ×2	8 ×4	6 ×7
5.	5 ×8	9 ×7	3 ×8	1 ×9	1 ×7	0 ×8
6.	5 ×9	8 ×0	3 ×9	5 ×7	8 ×9	2 ×7

Problem Solving

Solve each problem.

1. There are 8 chairs around each table. There are 9 tables. How many chairs are around all the tables?

   There are ________ chairs around each table.

   There are ________ tables.

   There are ________ chairs around all the tables.

2. Workers are eating lunch at 9 tables. Each table has 9 workers. How many workers are eating lunch?

   There are ________ tables.

   ________ workers are at each table.

   ________ workers are eating lunch.

3. The workers drink 9 liters of milk each day. They are at work 5 days each week. How many liters of milk do they drink in 5 days?

   They drink ________ liters of milk in 5 days.

4. A bowling league bowls 4 times each month. How many times will the league bowl in 9 months?

   The bowling league will bowl ________ times.

5. There are 9 packages of golf balls. Each package has 6 golf balls. How many golf balls are there in all?

   There are ________ golf balls.

6. A regular baseball game is 9 innings long. How many innings are in 7 regular games?

   There are ________ innings in 7 regular games.

1.	
2.	
3.	4.
5.	6.

NAME ______________________

Lesson 4 Multiplication

Multiply.

	a	*b*	*c*	*d*	*e*	*f*
1.	7×9	6×7	1×5	2×9	3×6	8×8
2.	3×7	4×9	0×8	7×5	7×8	6×6
3.	9×5	4×6	5×9	2×8	8×7	0×7
4.	7×6	1×6	9×8	0×9	5×5	9×7
5.	8×5	4×8	4×7	0×6	1×9	4×5
6.	6×8	9×6	6×5	6×4	7×7	3×9
7.	8×6	5×8	7×4	3×5	9×9	1×7
8.	9×4	5×7	1×8	8×9	5×6	2×5
9.	2×6	0×5	6×9	3×8	8×4	2×7

Problem Solving

Solve each problem.

1. Some students formed 5 teams. There were 8 students on each team. How many students were there?

There were __________ teams.

There were __________ students on each team.

There were __________ students in all.

1.

2. The waiter put 9 napkins on each table. There were 9 tables. How many napkins did the waiter use?

The waiter put __________ napkins on each table.

There were __________ tables.

The waiter used __________ napkins in all.

2.

3. Dr. Mede rides her bicycle 6 kilometers every day. How far would she ride in 9 days?

Dr. Mede rides __________ kilometers every day.

She rides for each of __________ days.

She would ride __________ kilometers in all.

4. Mr. Brown works 7 hours each day. How many hours will he work in 6 days?

Mr. Brown will work __________ hours in 6 days.

3.

4.

5. There are 8 hot dogs in each package. How many hot dogs are there in 9 packages?

There are __________ hot dogs in 9 packages.

6. Suppose you read 8 stories every day. How many stories would you read in 7 days?

You would read __________ stories in 7 days.

5.

6.

NAME ____________________

CHAPTER 8 TEST

Multiply.

	a	b	c	d	e
1.	6×7	5×9	8×6	4×7	7×7
2.	9×7	9×9	1×7	4×6	7×9
3.	0×6	7×8	5×8	9×8	7×6
4.	8×9	4×8	9×6	2×7	3×9

8

Solve each problem.

5. A clerk puts 6 oranges in each package. How many oranges are needed to make 9 packages?

There are __________ oranges in each package.

There are to be __________ packages.

__________ oranges are needed in all.

5.

6. A barbershop can handle 8 customers in one hour. How many customers can it handle in 8 hours?

It can handle __________ customers.

6.

7. Mr. Lawkin put 3 pictures in a row. He made 8 rows. How many pictures did he use?

Mr. Lawkin used __________ pictures.

7.

PRE-TEST—Multiplication

Multiply.

	a	*b*	*c*	*d*	*e*	*f*
1.	3 ×2	30 ×2	2 ×4	20 ×4	1 ×7	10 ×7
2.	32 ×3	24 ×2	13 ×3	21 ×4	11 ×5	23 ×3
3.	7 ×4	70 ×4	6 ×3	60 ×3	9 ×4	90 ×4
4.	62 ×3	74 ×2	31 ×8	62 ×4	83 ×2	41 ×5
5.	18 ×4	26 ×3	35 ×2	24 ×3	38 ×2	16 ×5
6.	43 ×4	27 ×5	35 ×3	46 ×6	28 ×6	54 ×5
7.	56 ×7	63 ×2	80 ×5	23 ×2	37 ×2	45 ×5
8.	17 ×4	40 ×8	73 ×3	22 ×3	54 ×3	27 ×6
9.	30 ×9	58 ×4	28 ×4	25 ×3	80 ×4	66 ×2

NAME ______________________

Lesson 1 Multiplication

4	40	2	20	7	70	8	80
×2	×2	×3	×3	×3	×3	×4	×4
8	80	6	60	21	210	32	320

Multiply.

	a	*b*	*c*	*d*	*e*	*f*
1.	3 ×2	30 ×2	2 ×4	20 ×4	6 ×1	60 ×1
2.	10 ×8	40 ×2	10 ×9	10 ×3	30 ×3	70 ×1
3.	9 ×4	90 ×4	6 ×3	60 ×3	5 ×5	50 ×5
4.	70 ×3	60 ×5	40 ×4	50 ×2	80 ×3	90 ×6
5.	20 ×4	30 ×5	40 ×3	20 ×2	30 ×6	40 ×7
6.	70 ×7	30 ×8	10 ×7	80 ×8	90 ×1	60 ×4
7.	40 ×5	20 ×8	60 ×2	50 ×3	10 ×5	40 ×1
8.	60 ×7	50 ×5	80 ×3	20 ×9	70 ×8	90 ×3

Problem Solving

Solve each problem.

1. There are 4 classrooms on the first floor. Each classroom has 30 seats. How many seats are on the first floor?

_________ seats are in each classroom.

_________ classrooms are on the first floor.

_________ seats are on the first floor.

1.

2. Laura placed 2 boxes on a wagon. Each box weighs 20 kilograms. How much do the two boxes weigh?

Each box weighs _________ kilograms.

Laura placed _________ boxes on the wagon.

The two boxes weigh _________ kilograms.

2.

3. Trent bought 3 packages of paper. Each package had 30 sheets. How many sheets of paper did he buy?

Each package contains _________ sheets of paper.

Trent bought _________ packages of paper.

Trent bought _________ sheets of paper.

4. Mrs. Long bought 6 boxes of nails. Each box had 30 nails. How many nails did she buy?

Mrs. Long bought _________ nails.

3.

4.

5. There are 40 bottles in a case. How many bottles are there in 2 cases?

There are _________ bottles in 2 cases.

6. A bus has 40 passenger seats. How many passenger seats would there be on 7 such buses?

There are _________ passenger seats on 7 buses.

5.

6.

NAME ______________________

Lesson 2 Multiplication

	Multiply 2 ones by 3.	Multiply 7 tens by 3.	
72 × 3	72 × 3 6	72 × 3 6 210	72 × 3 6 210 } Add. 216

Multiply.

	a	*b*	*c*	*d*	*e*	*f*
1.	62 ×2	73 ×3	92 ×4	84 ×2	53 ×2	42 ×3
2.	23 ×2	31 ×3	42 ×2	12 ×4	33 ×3	42 ×2
3.	21 ×4	61 ×3	51 ×2	43 ×2	82 ×4	13 ×3
4.	72 ×3	32 ×3	43 ×3	23 ×3	34 ×2	93 ×2

NAME ______________________________

Lesson 3 Multiplication

	Multiply 4 ones by 2.	Multiply. 8 tens by 2.
$\begin{array}{r} 84 \\ \times 2 \\ \hline \end{array}$	$\begin{array}{r} 84 \\ \times 2 \\ \hline 8 \end{array}$	$\begin{array}{r} 84 \\ \times 2 \\ \hline 168 \end{array}$

Multiply.

	a	*b*	*c*	*d*	*e*	*f*
1.	62×4	84×2	73×3	82×2	91×2	52×3
2.	43×2	41×4	32×3	61×3	42×4	44×2
3.	72×3	71×3	81×2	43×3	52×4	83×2
4.	62×2	93×3	52×1	74×2	53×2	62×3
5.	72×4	81×5	92×3	63×3	54×2	73×2
6.	61×7	32×4	82×3	81×9	63×2	91×5
7.	53×3	71×6	82×4	91×9	92×4	81×8

NAME ____________________

Lesson 4 Multiplication

	Multiply 7 ones by 3.	Multiply 1 ten by 3. Add the 2 tens.	
17 ×3	2 17 ×3 1	2 17 ×3 51	17 ×3 51
	3×7 = 21 = 20+1	3 × 10 = 30 30 + 20 = 50	

Multiply.

	a	b	c	d	e	f
1.	23 ×4	29 ×3	16 ×5	14 ×7	26 ×3	12 ×8
2.	37 ×2	26 ×2	47 ×2	28 ×3	15 ×5	24 ×4
3.	18 ×5	14 ×5	28 ×3	35 ×2	46 ×2	38 ×2
4.	45 ×2	27 ×3	15 ×6	12 ×7	15 ×4	48 ×2
5.	28 ×2	12 ×6	17 ×5	13 ×6	19 ×3	19 ×4
6.	36 ×2	24 ×3	25 ×3	16 ×4	29 ×2	18 ×3

Problem Solving

Solve each problem.

1. Ms. McClean ordered 7 dozen radio antennas. How many antennas did she order? (There are 12 items in a dozen.)

There are _________ items in a dozen.

She ordered _________ dozen antennas.

Ms. McClean ordered _________ radio antennas.

2. There are 14 CD players on a shelf. Each player weighs 5 kilograms. How much do all the players weigh?

There are _________ CD players.

Each player weighs _________ kilograms.

All the players weigh _________ kilograms.

3. Mr. Tunin bought 2 CD players. Each player cost $49. How much did both players cost?

Both players cost $ _________.

4. Ms. McClean sold 36 radios this week. She sold the same number of radios last week. How many radios did she sell in the two weeks?

She sold _________ radios in the two weeks.

1.

2.

3.

4.

NAME ____________________

Lesson 5 Multiplication

Multiply 7 ones by 5.

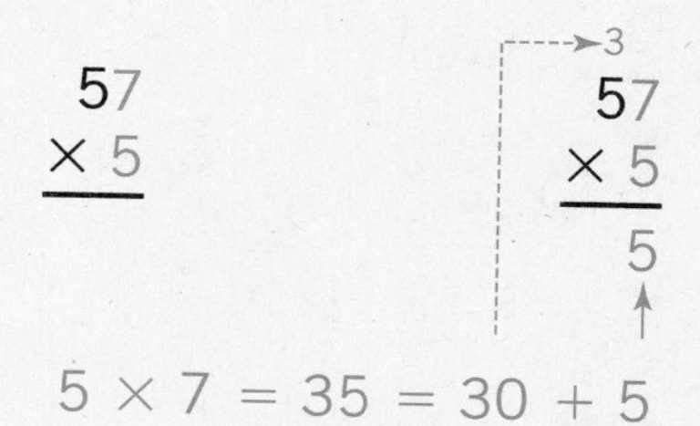

Multiply 5 tens by 5. Add the 3 tens.

$$5 \times 50 = 250$$

$$250 + 30 = 280 = 200 + 80$$

$$\begin{array}{r} {}^{3} \\ 57 \\ \times\ 5 \\ \hline 285 \end{array}$$

Multiply.

	a	*b*	*c*	*d*	*e*	*f*
1.	35 ×4	42 ×6	56 ×3	47 ×5	38 ×6	25 ×5
2.	54 ×4	27 ×6	38 ×5	48 ×8	83 ×7	74 ×6
3.	75 ×4	58 ×3	46 ×4	37 ×6	29 ×5	46 ×3
4.	84 ×4	93 ×6	62 ×8	57 ×5	39 ×4	22 ×7
5.	45 ×6	68 ×7	73 ×9	87 ×8	94 ×6	83 ×4
6.	96 ×5	85 ×3	47 ×4	23 ×9	39 ×7	65 ×6

Problem Solving

Solve each problem.

1. Jordan's spelling book has 25 new words on each page. There are 9 pages in the first section. How many new words are there in the first section?

There are ______________ new spelling words on each page.

There are ________ pages in the first section.

There are ______________ new spelling words in the first section.

2. Alan wants to walk up 6 flights of stairs. There are 26 steps in each flight. How many steps will he have to walk up?

There are ______________ steps in each flight.

Alan wants to walk up ______________ flights.

Alan will have to walk up ______________ steps.

3. There are 7 rows of seats in the balcony. There are 36 seats in each row. How many seats are in the balcony?

There are ______________ seats in each row.

There are ______________ rows.

There are ______________ seats in the balcony.

4. There are 25 baseball players on each team. How many players are there on 8 such teams?

There are ______________ players on 8 teams.

5. Brenna used 3 rolls of film. She took 36 pictures on each roll. How many pictures did Brenna take?

Brenna took ______________ pictures.

1.

2.

3.

4.

5.

NAME ______________________

CHAPTER 9 TEST

Multiply.

	a	*b*	*c*	*d*	*e*
1.	30×2	42×2	23×3	60×4	80×3
2.	84×2	73×3	21×7	14×6	27×3
3.	57×5	38×6	42×5	29×4	36×5
4.	15×4	73×2	58×3	40×9	28×3

9

Solve each problem.

5. Morgan has 4 decks of cards. There are 52 cards in each deck. How many cards does she have?

She has ______________ decks.

There are ______________ cards in each deck.

Morgan has ______________ cards in all.

6. Blake's father works 37 hours each week. How many hours would he work in 4 weeks?

He would work ______________ hours in 4 weeks.

7. Mr. Richards gave each student 3 sheets of paper. There were 28 students. How many sheets of paper did he use?

Mr. Richards used ______________ sheets of paper.

5.

6.

7.

NAME ______________________

PRE-TEST—Division

Divide.

	a	*b*	*c*	*d*	*e*
1.	$2\overline{)6}$	$2\overline{)12}$	$2\overline{)18}$	$2\overline{)4}$	$2\overline{)10}$
2.	$4\overline{)16}$	$3\overline{)24}$	$3\overline{)9}$	$5\overline{)25}$	$3\overline{)3}$
3.	$1\overline{)4}$	$1\overline{)8}$	$1\overline{)16}$	$4\overline{)36}$	$1\overline{)27}$
4.	$2\overline{)16}$	$1\overline{)1}$	$3\overline{)27}$	$1\overline{)6}$	$1\overline{)21}$
5.	$1\overline{)12}$	$5\overline{)20}$	$2\overline{)14}$	$4\overline{)8}$	$1\overline{)14}$
6.	$1\overline{)5}$	$4\overline{)28}$	$5\overline{)5}$	$1\overline{)2}$	$3\overline{)18}$

Solve each problem.

7. Paula has 18 books. She put them in piles of 3 books each. How many piles of books does she have?

She has __________ piles of books.

7.

8. Bart has 12 pennies. He put 3 pennies in each stack. How many stacks of pennies does he have?

He has __________ stacks of pennies.

8.

9. Dee has 12 pennies. She put 2 pennies in each stack. How many stacks of pennies does she have?

She has __________ stacks of pennies.

9.

NAME ______________________

Lesson 1 Division

÷ and $\overline{)}$ mean divide.

6 ÷ 2 = 3 is read "6 divided by 2 is equal to 3."

8 ÷ 2 = 4 is read "_____ divided by 2 is equal to _____."

$2\overline{)6}^{\,3}$ is read "6 divided by 2 is equal to 3."

$2\overline{)8}^{\,4}$ is read "_____ divided by 2 is equal to _____."

divisor ------→ $2\overline{)6}$ ←------ dividend; 3 ←------ quotient

In $2\overline{)8}^{\,4}$, the divisor is _____, the dividend is _____, and the quotient is _____.

Complete each sentence.

1. 10 ÷ 2 = 5 is read "__________ divided by 2 is equal to __________."
2. 21 ÷ 3 = 7 is read "__________ divided by 3 is equal to __________."
3. 4 ÷ 2 = 2 is read "__________ divided by 2 is equal to __________."
4. $3\overline{)18}^{\,6}$ is read "__________ divided by 3 is equal to __________."
5. $2\overline{)18}^{\,9}$ is read "__________ divided by 2 is equal to __________."
6. $3\overline{)24}^{\,8}$ is read "__________ divided by 3 is equal to __________."
7. In $3\overline{)21}^{\,7}$, the divisor is ________, the dividend is ________, and the quotient is ________.
8. In $2\overline{)4}^{\,2}$, the divisor is ________, the dividend is ________, and the quotient is ________.
9. In $2\overline{)10}^{\,5}$, the divisor is ________, the dividend is ________, and the quotient is ________.
10. In $3\overline{)18}^{\,6}$, the divisor is ________, the dividend is ________, and the quotient is ________.

Lesson 2 Division

6 ×'s in all.
2 ×'s in each group.

How many groups?

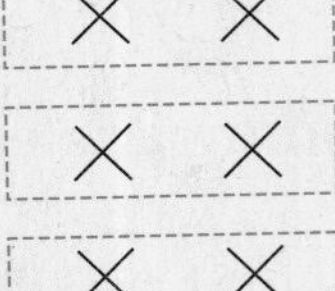

6 ÷ 2 = 3

There are 3 groups.

6 ×'s in all.
3 groups of ×'s.

How many ×'s in each group?

6 ÷ 3 = ____

There are ____ ×'s in each group.

Complete the following.

a

1. 10 ☆'s in all.
2 ☆'s in each group.

How many groups?

10 ÷ 2 = ____

There are ____ groups.

b

10 ☆'s in all.
5 groups of ☆'s.

How many ☆'s in each group?

10 ÷ 5 = ____

There are ____ ☆'s in each group.

2. 8 □'s in all.

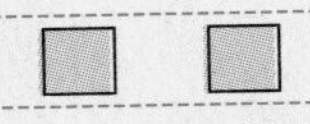

____ □'s in each group.

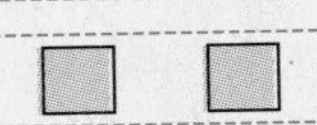

How many groups?

8 ÷ 2 = ____

There are ____ groups.

____ □'s in all.

4 groups of □'s.
How many □'s in each group?

8 ÷ 4 = ____

There are ____ □'s in each group.

3. ____ ○'s in all.

____ ○'s in each group.
How many groups?

4 ÷ 2 = ____

There are ____ groups.

____ ○'s in all.

____ groups of ○'s.
How many ○'s in each group?

4 ÷ 2 = ____

There are ____ ○'s in each group.

NAME ____________________

Lesson 3 Division

$$\begin{array}{r} 3 \\ \times 2 \\ \hline 6 \end{array} \dashrightarrow 2\overline{)\,6}^{\,3}$$

$$\begin{array}{r} 4 \\ \times 3 \\ \hline 12 \end{array} \dashrightarrow 3\overline{)\,12}^{\,4}$$

If $2 \times 3 = 6$, then $6 \div 2 = 3$.

If $3 \times 4 = 12$, then ________ $\div 3 =$ ________.

Divide as shown.

	a		*b*	
1.	$\begin{array}{r} 5 \\ \times 2 \\ \hline 10 \end{array}$	$2\overline{)\,1\,0}$ (5)	$\begin{array}{r} 6 \\ \times 3 \\ \hline 18 \end{array}$	$3\overline{)\,1\,8}$
2.	$\begin{array}{r} 7 \\ \times 2 \\ \hline 14 \end{array}$	$2\overline{)\,1\,4}$	$\begin{array}{r} 8 \\ \times 3 \\ \hline 24 \end{array}$	$3\overline{)\,2\,4}$
3.	$\begin{array}{r} 1 \\ \times 2 \\ \hline 2 \end{array}$	$2\overline{)\,2}$	$\begin{array}{r} 3 \\ \times 3 \\ \hline 9 \end{array}$	$3\overline{)\,9}$
4.	$\begin{array}{r} 8 \\ \times 2 \\ \hline 16 \end{array}$	$2\overline{)\,1\,6}$	$\begin{array}{r} 9 \\ \times 3 \\ \hline 27 \end{array}$	$3\overline{)\,2\,7}$

Divide.

	a	*b*	*c*	*d*
5.	$3\overline{)\,1\,5}$	$2\overline{)\,6}$	$3\overline{)\,3}$	$3\overline{)\,6}$
6.	$3\overline{)\,2\,1}$	$2\overline{)\,1\,8}$	$2\overline{)\,8}$	$2\overline{)\,1\,2}$

Problem Solving

Solve each problem.

1. Twenty-four people are at work. They work in 3 departments. The same number of people work in each department. How many people work in each department?

   There are __________ people.

   They work in __________ departments.

   There are __________ people in each department.

2. Dan put 8 books into 2 stacks. Each stack had the same number of books. How many books were in each stack?

   There were __________ books in all.

   They were put into __________ stacks.

   There were __________ books in each stack.

3. Janice put 16 liters of water into 2 jars. She put the same number of liters into each jar. How many liters of water did she put into each jar?

   Janice put __________ liters of water into jars.

   She used __________ jars.

   Janice put __________ liters of water into each jar.

4. Kim has 27 apples. She wants to put the same number of apples in each of 3 boxes. How many apples should she put in each box?

   She should put __________ apples in each box.

5. Mr. Green had 18 inches of wire. He cut the wire into 2 pieces. The pieces were the same length. How long was each piece?

   Each piece was __________ inches long.

1.

2.

3.

4.

5.

NAME ________________

Lesson 4 Division

$$\begin{array}{r} 5 \\ \times 4 \\ \hline 20 \end{array} \quad 4\overline{)20} = 5$$

$$\begin{array}{r} 9 \\ \times 5 \\ \hline 45 \end{array} \quad 5\overline{)45} = 9$$

If 4 × 5 = 20, then 20 ÷ 4 = 5. If 5 × 9 = 45, then ________ ÷ 5 = ________.

Divide as shown.

	a		*b*	
1.	7 × 4 = 28	$4\overline{)28}$	6 × 5 = 30	$5\overline{)30}$
2.	4 × 4 = 16	$4\overline{)16}$	3 × 5 = 15	$5\overline{)15}$
3.	6 × 4 = 24	$4\overline{)24}$	4 × 5 = 20	$5\overline{)20}$
4.	9 × 4 = 36	$4\overline{)36}$	8 × 5 = 40	$5\overline{)40}$

Divide.

	a	*b*	*c*	*d*
5.	$4\overline{)8}$	$5\overline{)10}$	$4\overline{)4}$	$4\overline{)12}$
6.	$5\overline{)25}$	$5\overline{)5}$	$4\overline{)32}$	$5\overline{)35}$

Problem Solving

Solve each problem.

1. A loaf of bread has 24 slices. Mrs. Spencer uses 4 slices each day. How long will a loaf of bread last her?

A loaf of bread has __________ slices.

Mrs. Spencer uses __________ slices a day.

The loaf of bread will last __________ days.

1.

2. A football team played 28 periods. There are 4 periods in a game. How many games did they play?

The football team played __________ periods.

There are __________ periods each game.

The football team played __________ games.

2.

3. A basketball game is 32 minutes long. The game is separated into 4 parts. Each part has the same number of minutes. How long is each part?

A basketball game is __________ minutes long.

The game is separated into __________ parts.

Each part is __________ minutes long.

3.

4. Emma worked 25 problems. She worked 5 problems on each sheet of paper. How many sheets of paper did she use?

She used __________ sheets of paper.

4.

5. Robert works the same number of hours each week. He worked 45 hours in 5 weeks. How many hours does he work each week?

Robert works __________ hours each week.

5.

NAME ____________________

Lesson 5 Division

$\begin{array}{r} 8 \\ \times 1 \\ \hline 8 \end{array}$ → $1\overline{)8}$	$\begin{array}{r} 15 \\ \times 1 \\ \hline 15 \end{array}$ → $1\overline{)15}$
If 1 × 8 = 8, then 8 ÷ 1 = 8.	If 1 × 15 = 15, then ______ ÷ 1 = ______.

Divide.

	a		*b*	
1.	$\begin{array}{r} 5 \\ \times 1 \\ \hline 5 \end{array}$	$1\overline{)5}$	$\begin{array}{r} 14 \\ \times 1 \\ \hline 14 \end{array}$	$1\overline{)14}$
2.	$\begin{array}{r} 4 \\ \times 1 \\ \hline 4 \end{array}$	$1\overline{)4}$	$\begin{array}{r} 9 \\ \times 1 \\ \hline 9 \end{array}$	$1\overline{)9}$

	a	*b*	*c*	*d*	*e*
3.	$1\overline{)4}$	$1\overline{)3}$	$1\overline{)12}$	$1\overline{)2}$	$1\overline{)16}$
4.	$2\overline{)8}$	$3\overline{)18}$	$2\overline{)18}$	$2\overline{)6}$	$3\overline{)6}$
5.	$4\overline{)16}$	$2\overline{)14}$	$1\overline{)9}$	$5\overline{)5}$	$5\overline{)45}$
6.	$2\overline{)16}$	$4\overline{)12}$	$2\overline{)10}$	$4\overline{)28}$	$1\overline{)18}$
7.	$4\overline{)4}$	$4\overline{)20}$	$5\overline{)10}$	$5\overline{)30}$	$4\overline{)32}$

Problem Solving

Solve each problem.

1. Dana bought 16 rolls. The rolls came in 2 packs. The same number of rolls were in each pack. How many rolls were in each pack? Dana bought _________ rolls. These rolls filled _________ packs. There were _________ rolls in each pack.	**1.**
2. There are 9 families in an apartment building. There are 3 families on each floor. How many floors are in the building? There are _________ families in the building. There are _________ families on each floor. There are _________ floors in the building.	**2.**
3. Arlene put 36 oranges in bags. She put 4 oranges in each bag. How many bags did she fill? Arlene put _________ oranges in bags. She put _________ oranges in each bag. Arlene filled _________ bags with oranges.	**3.**
4. Marcos read 35 pages of science in 5 days. He read the same number of pages each day. How many pages did he read each day? Marcos read _________ pages each day.	**4.**
5. Mrs. Allan worked 25 hours in 5 days. She worked the same number of hours each day. How many hours did she work each day? Mrs. Allan worked _________ hours each day.	**5.**

NAME ________________

CHAPTER 10 TEST

Divide.

	a	b	c	d	e
1.	$2\overline{)10}$	$1\overline{)4}$	$3\overline{)3}$	$3\overline{)9}$	$2\overline{)16}$
2.	$1\overline{)12}$	$2\overline{)12}$	$3\overline{)12}$	$2\overline{)14}$	$3\overline{)15}$
3.	$3\overline{)6}$	$1\overline{)8}$	$5\overline{)20}$	$1\overline{)9}$	$3\overline{)24}$
4.	$5\overline{)40}$	$5\overline{)5}$	$1\overline{)10}$	$4\overline{)36}$	$4\overline{)24}$

Solve each problem.

5. The 45 students in a class separated into 5 groups. Each group has the same number of students. How many are in each group?

There are __________ students in all.

The students are separated into __________ groups.

There are __________ students in each group.

5.

6. Sydney has 28 balloons for a party. She will give each person 4 balloons. How many people will receive balloons?

__________ people will receive balloons.

6.

7. Mr. Graham has 6 birds. How many cages does he need in order to put 2 birds in each cage?

Mr. Graham needs __________ cages.

7.

10

PRE-TEST—Division

Divide.

	a	*b*	*c*	*d*	*e*
1.	$6\overline{)24}$	$6\overline{)12}$	$6\overline{)18}$	$6\overline{)0}$	$6\overline{)6}$
2.	$6\overline{)42}$	$6\overline{)54}$	$6\overline{)30}$	$6\overline{)36}$	$6\overline{)48}$
3.	$7\overline{)0}$	$7\overline{)28}$	$7\overline{)14}$	$7\overline{)21}$	$7\overline{)7}$
4.	$7\overline{)56}$	$7\overline{)42}$	$7\overline{)63}$	$7\overline{)35}$	$7\overline{)49}$
5.	$8\overline{)8}$	$8\overline{)40}$	$8\overline{)0}$	$8\overline{)32}$	$8\overline{)16}$
6.	$8\overline{)24}$	$8\overline{)48}$	$8\overline{)64}$	$8\overline{)72}$	$8\overline{)56}$
7.	$9\overline{)36}$	$9\overline{)27}$	$9\overline{)45}$	$9\overline{)18}$	$9\overline{)0}$
8.	$9\overline{)72}$	$9\overline{)63}$	$9\overline{)54}$	$9\overline{)9}$	$9\overline{)81}$
9.	$5\overline{)5}$	$4\overline{)28}$	$1\overline{)1}$	$5\overline{)30}$	$4\overline{)12}$

NAME ____________________

Lesson 1 Division

$$\begin{array}{r} 3 \\ \times 6 \\ \hline 18 \end{array} \quad 6\overline{)18} = 3$$

If 6 × 3 = 18, then 18 ÷ 6 = 3.

$$\begin{array}{r} 4 \\ \times 6 \\ \hline 24 \end{array} \quad 6\overline{)24} = 4$$

If 6 × 4 = 24, then ________ ÷ 6 = ________.

Divide.

	a		*b*	
1.	$\begin{array}{r} 2 \\ \times 6 \\ \hline 12 \end{array}$	$6\overline{)12}$	$\begin{array}{r} 1 \\ \times 6 \\ \hline 6 \end{array}$	$6\overline{)6}$
2.	$\begin{array}{r} 5 \\ \times 6 \\ \hline 30 \end{array}$	$6\overline{)30}$	$\begin{array}{r} 7 \\ \times 6 \\ \hline 42 \end{array}$	$6\overline{)42}$
3.	$\begin{array}{r} 8 \\ \times 6 \\ \hline 48 \end{array}$	$6\overline{)48}$	$\begin{array}{r} 9 \\ \times 6 \\ \hline 54 \end{array}$	$6\overline{)54}$

	a	*b*	*c*	*d*
4.	$6\overline{)6}$	$6\overline{)12}$	$6\overline{)36}$	$6\overline{)18}$
5.	$1\overline{)6}$	$6\overline{)0}$	$6\overline{)24}$	$6\overline{)42}$
6.	$6\overline{)30}$	$6\overline{)54}$	$6\overline{)48}$	$5\overline{)45}$
7.	$4\overline{)32}$	$5\overline{)20}$	$4\overline{)20}$	$5\overline{)30}$

Problem Solving

Solve each problem.

1. There are 6 rows of mailboxes. Each row has the same number of mailboxes. There are 30 mailboxes in all. How many are in each row?

There are __________ mailboxes in all.

The mailboxes are separated into __________ rows.

There are __________ mailboxes in each row.

1.

2. The movie was shown 12 times in 6 days. It was shown the same number of times each day. How many times was it shown each day?

The movie was shown __________ times in all.

The movie was shown for __________ days.

The movie was shown __________ times each day.

2.

3. Jill bought 18 buttons. The buttons were on cards of 6 buttons each. How many cards were there?

Jill bought __________ buttons.

There were __________ buttons on a card.

There were __________ cards.

3.

4. Spencer got 6 hits in 6 games. He got the same number of hits in each game. How many hits did he get in each game?

Spencer got __________ hit in each game.

4.

5. One side of a building has 24 windows. Each floor has 6 windows on that side. How many floors does the building have?

The building has __________ floors.

5.

NAME ____________________

Lesson 2 Division

$$\begin{array}{r} 3 \\ \times 7 \\ \hline 21 \end{array} \qquad 7\overline{)21} = 3$$

If 7 × 3 = 21, then 21 ÷ 7 = 3.

$$\begin{array}{r} 5 \\ \times 8 \\ \hline 40 \end{array} \qquad 8\overline{)40} = 5$$

If 8 × 5 = 40, then ________ ÷ 8 = ________.

Divide.

	a		*b*	
1.	$\begin{array}{r} 2 \\ \times 7 \\ \hline 14 \end{array}$	$7\overline{)14}$	$\begin{array}{r} 3 \\ \times 8 \\ \hline 24 \end{array}$	$8\overline{)24}$
2.	$\begin{array}{r} 5 \\ \times 7 \\ \hline 35 \end{array}$	$7\overline{)35}$	$\begin{array}{r} 4 \\ \times 8 \\ \hline 32 \end{array}$	$8\overline{)32}$
3.	$\begin{array}{r} 7 \\ \times 7 \\ \hline 49 \end{array}$	$7\overline{)49}$	$\begin{array}{r} 8 \\ \times 8 \\ \hline 64 \end{array}$	$8\overline{)64}$

	a	*b*	*c*	*d*
4.	$7\overline{)7}$	$8\overline{)0}$	$8\overline{)16}$	$7\overline{)28}$
5.	$8\overline{)48}$	$7\overline{)42}$	$8\overline{)8}$	$7\overline{)56}$
6.	$7\overline{)0}$	$8\overline{)56}$	$1\overline{)7}$	$7\overline{)63}$
7.	$1\overline{)8}$	$8\overline{)40}$	$8\overline{)72}$	$7\overline{)21}$

Problem Solving

Solve each problem.

1. A classroom has 28 chairs in 7 rows. Each row has the same number of chairs. How many chairs are in each row?

   There are __________ chairs in the classroom.

   The chairs are separated into __________ rows.

   There are __________ chairs in each row.

2. There are 48 chairs around the tables in the library. There are 8 chairs for each table. How many tables are in the library?

   There are __________ chairs in the library.

   There are __________ chairs around each table.

   There are __________ tables in the library.

3. Zane worked the same number of hours each day. He worked 21 hours in 7 days. How many hours did he work each day?

   Zane worked __________ hours each day.

4. There are 16 cars in the parking lot. There are 8 cars in each row. How many rows of cars are there?

   There are __________ rows of cars.

5. Mr. Miller sold 7 cars in 7 days. He sold the same number of cars each day. How many did he sell each day?

   Mr. Miller sold __________ car each day.

1.
2.
3.
4.
5.

NAME ______________________

Lesson 3 Division

$$\begin{array}{r} 2 \\ \times 9 \\ \hline 18 \end{array} \quad 9\overline{)18} = 2$$

If 9 × 2 = 18, then 18 ÷ 9 = 2.

$$\begin{array}{r} 7 \\ \times 9 \\ \hline 63 \end{array} \quad 9\overline{)63} = 7$$

If 9 × 7 = 63, then ______ ÷ 9 = ______.

Divide.

	a		b	
1.	5 × 9 = 45	$9\overline{)45}$	3 × 9 = 27	$9\overline{)27}$
2.	8 × 9 = 72	$9\overline{)72}$	4 × 9 = 36	$9\overline{)36}$
3.	6 × 9 = 54	$9\overline{)54}$	9 × 9 = 81	$9\overline{)81}$

	a	b	c	d
4.	$9\overline{)9}$	$1\overline{)9}$	$9\overline{)18}$	$9\overline{)36}$
5.	$9\overline{)0}$	$9\overline{)72}$	$9\overline{)54}$	$9\overline{)81}$
6.	$8\overline{)72}$	$9\overline{)63}$	$8\overline{)48}$	$9\overline{)45}$
7.	$9\overline{)27}$	$8\overline{)56}$	$7\overline{)63}$	$7\overline{)49}$

Problem Solving

Solve each problem.

1. A farmer planted 54 cherry trees in 9 rows. Each row had the same number of trees. How many trees were in each row?

A farmer planted __________ trees.

There were __________ rows of trees.

There were __________ trees in each row.

1.

2. Curt put 27 tennis balls in 9 cans. He put the same number of balls in each can. How many balls did Curt put in each can?

Curt put __________ tennis balls in cans.

There were __________ cans.

He put __________ balls in each can.

2.

3. There are 9 packs of batteries on a shelf. Each pack has the same number of batteries. There are 36 batteries in all. How many batteries are in each pack?

There are __________ batteries in each pack.

3.

4. There are 18 cornstalks in a garden. There are 9 stalks in each row. How many rows of cornstalks are there?

There are __________ rows of cornstalks.

4.

5. Kay had 45 pennies. She put the pennies into stacks of 9 pennies each. How many stacks of pennies did she make?

She made __________ stacks of pennies.

5.

NAME ______________________________

Lesson 4 Division

Divide.

	a	*b*	*c*	*d*
1.	$2\overline{)10}$	$3\overline{)18}$	$4\overline{)4}$	$1\overline{)8}$
2.	$5\overline{)15}$	$8\overline{)16}$	$6\overline{)24}$	$7\overline{)42}$
3.	$2\overline{)18}$	$3\overline{)24}$	$7\overline{)35}$	$9\overline{)0}$
4.	$5\overline{)25}$	$4\overline{)32}$	$9\overline{)27}$	$6\overline{)36}$
5.	$7\overline{)14}$	$3\overline{)15}$	$8\overline{)8}$	$2\overline{)16}$
6.	$9\overline{)18}$	$6\overline{)12}$	$3\overline{)12}$	$8\overline{)24}$
7.	$5\overline{)20}$	$4\overline{)12}$	$2\overline{)6}$	$5\overline{)10}$
8.	$7\overline{)56}$	$3\overline{)21}$	$8\overline{)40}$	$6\overline{)30}$
9.	$4\overline{)28}$	$9\overline{)45}$	$7\overline{)49}$	$9\overline{)72}$
10.	$8\overline{)64}$	$9\overline{)54}$	$8\overline{)48}$	$9\overline{)81}$

Problem Solving

Solve each problem.

1. Olivia has 42 apples. She puts 6 apples in a package. How many packages will she have?

   Olivia has __________ apples.

   She puts __________ apples in each package.

   There will be __________ packages of apples.

2. Olivia has 63 peaches. She puts 7 peaches in a package. How many packages will she have?

   Olivia has __________ peaches.

   Each package will have __________ peaches.

   There will be __________ packages of peaches.

3. There are 8 packages of pears. Each package has the same number of pears. There are 64 pears in all. How many pears are in each package?

   There are __________ pears in all.

   There are __________ packages of pears.

   There are __________ pears in each package.

1.

2.

3.

NAME ______________________

CHAPTER 11 TEST

Divide.

	a	*b*	*c*	*d*	*e*
1.	6)12	7)7	8)24	6)36	9)0
2.	7)14	9)45	6)42	8)32	1)9
3.	6)48	7)21	8)40	9)18	8)72
4.	8)64	9)81	7)56	6)54	6)18
5.	6)30	7)28	9)72	7)63	8)48

Solve each problem.

6. A classroom has 24 desks. They are in 6 rows. There is the same number of desks in each row. How many desks are in each row?

There are __________ desks in all.

There are __________ rows with the same number of desks in each row.

There are __________ desks in each row.

6.

7. Lyle put 24 biscuits on a tray. He put 8 biscuits in each row. How many rows were there?

There were __________ rows.

7.

11

NAME ______________________

Chapter 12

PRE-TEST—Metric Measurement

Find the length of each object to the nearest centimeter.

1. ________ centimeters

2. ________ centimeters

3. ________ centimeters

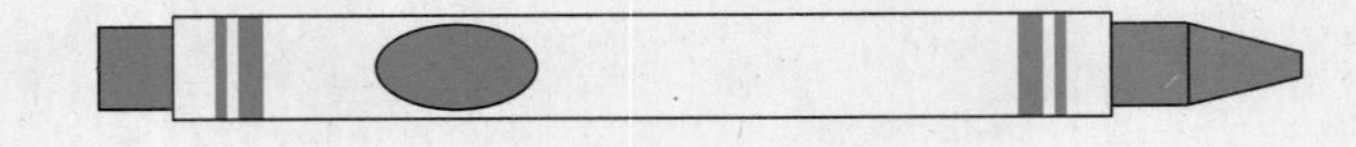

4. ________ centimeters

5. ________ centimeters

12

Answer *True* or *False*.

6. A liter is less than 800 milliliters. ____________________

7. A liter is more than 1800 milliliters. ____________________

8. A liter is equal to 1000 milliliters. ____________________

Solve.

9. A car can go 6 kilometers on a liter of gasoline. The car has a tank that holds 55 liters. How far can the car go on a full tank of gasoline?

The car can go __________ kilometers.

NAME ____________________

Lesson 1 Centimeter

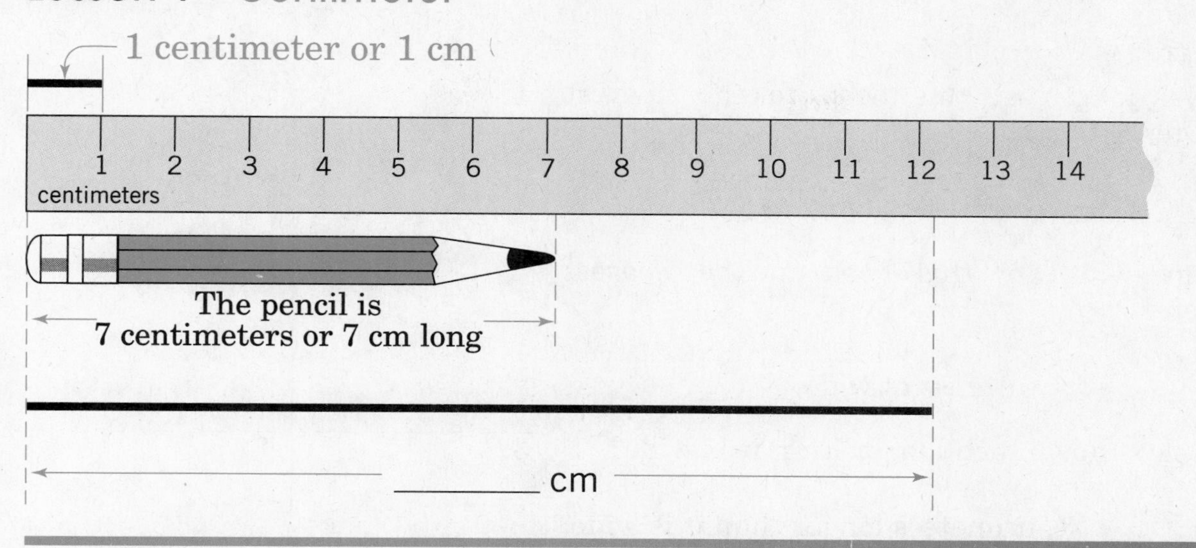

Estimate how long each object is in centimeters.
Then find the length of each object to the nearest centimeter.

1. Estimate: _______ cm 

Length: _______ cm

2. Estimate: _______ cm 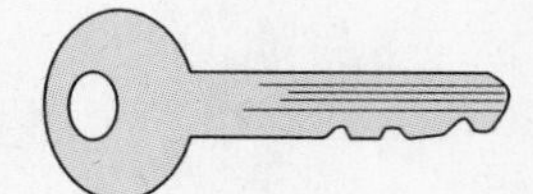

Length: _______ cm

3. Estimate: _______ cm 

Length: _______ cm

4. Estimate: _______ cm 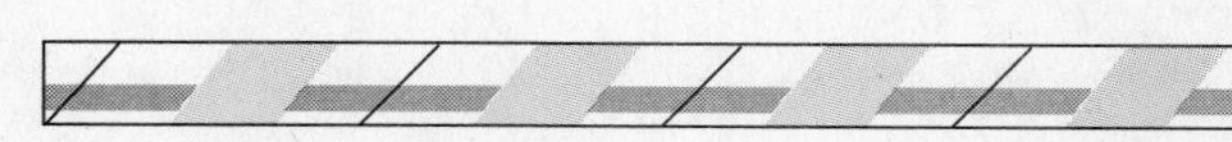

Length: _______ cm

5. Estimate: _______ cm

Length: _______ cm

6. Estimate: _______ cm

Length: _______ cm

Problem Solving

Solve each problem.

1. Find the length of this book to the nearest centimeter.

It is __________ centimeters long.

2. Find the width of this book to the nearest centimeter.

It is __________ centimeters wide.

3. This book is how much longer than it is wide?

It is __________ centimeters longer than it is wide.

4. How many centimeters is it across a nickel?

It is __________ centimeters across.

5. How many centimeters would it be across 8 nickels laid in a row?

It would be __________ centimeters across.

6. Find the length of your shoe to the nearest centimeter.

It is __________ centimeters long.

1.	**2.**
3.	**4.**
5.	**6.**

Use a tape measure or string to find the following to the nearest centimeter.

7. the distance around your wrist __________ centimeters

8. the distance around your waist __________ centimeters

9. the distance around your head __________ centimeters

10. the distance around your ankle __________ centimeters

NAME ________________

Lesson 2 Metric Measurement

From A to B is a length of 12 centimeters.

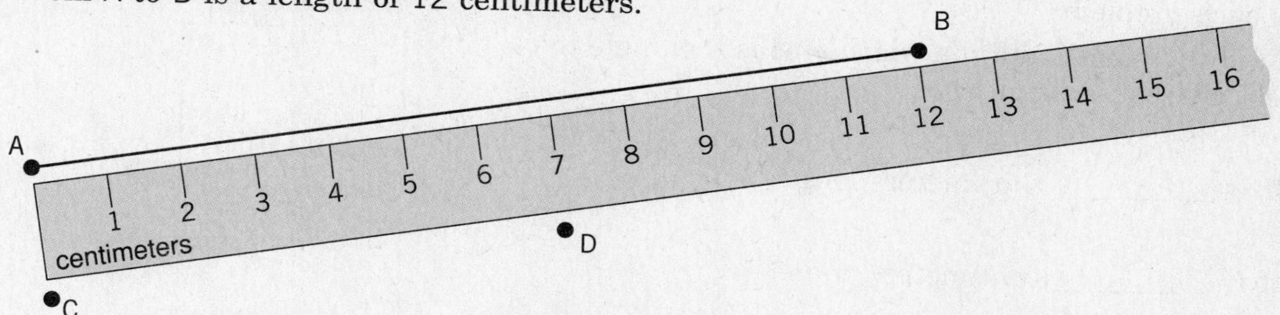

Draw from C to D. It is ______ centimeters long.

1. Draw from E to F.

   The length is ________ centimeters.

2. Draw from G to H.

   The length is ________ centimeters.

3. Draw from J to K.

   The length is ________ centimeters.

E H G K J F

Complete the table.

	From	*Length*
4.	A to B	________ cm
5.	B to C	________ cm
6.	C to D	________ cm
7.	D to E	________ cm
8.	E to A	________ cm
9.	A to D	________ cm
10.	C to E	________ cm

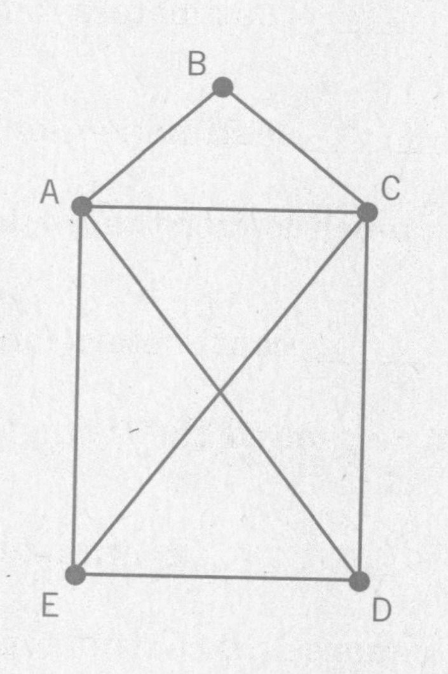

Problem Solving

Solve each problem.

1. Find the length and the width of this rectangle to the nearest centimeter.

It is _________ centimeters long.

It is _________ centimeters wide.

2. The rectangle is how much longer than it is wide?

It is _________ centimeters longer than it is wide.

3. Find the distance around the rectangle.

The distance is _________ centimeters.

4. Draw from A to B, from B to C, and from C to A. Then find the length of each side of the triangle you just drew.

A •

• B

C •

Side AB is _________ centimeters long.

Side BC is _________ centimeters long.

Side CA is _________ centimeters long.

5. Side CA is how much longer than side BC?

Side CA is _________ centimeters longer.

6. Find the distance around the triangle.

The distance is _________ centimeters.

7. One side of a square is 8 centimeters long. What is the distance around the square? (All 4 sides of a square are the same length.)

The distance is _________ centimeters.

NAME ____________________

Lesson 3 Liter

1 liter is a little more than 1 quart.

Answer *Yes* or *No*.

1. You can put 1 quart of water in a 1-liter bottle. __________

2. You can put 3 liters of water in a 3-quart pail. __________

How many liters would each container hold?
Underline the best answer.

3.

1 liter 8 liters 45 liters

4.

2 liters 9 liters 25 liters

5.

1 liter 10 liters 50 liters

6.

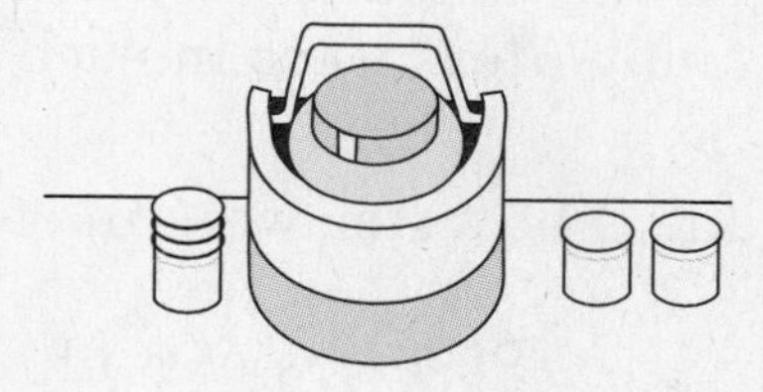

8 liters 28 liters 64 liters

7.

4 liters 16 liters 80 liters

8.

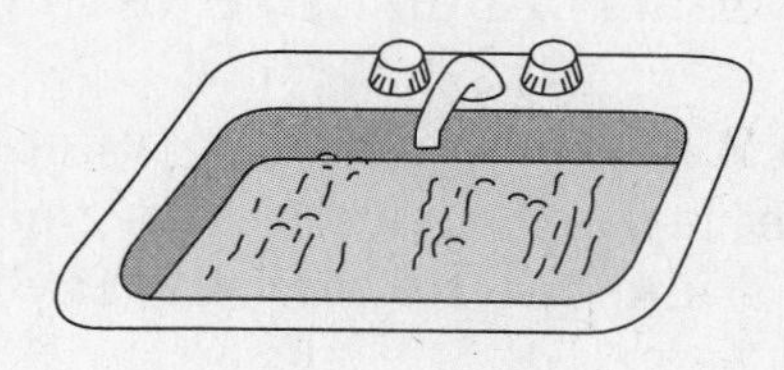

1 liter 4 liters 20 liters

Problem Solving

Solve each problem.

1. The tank in Mr. Sumner's car can hold 85 liters. It took 37 liters of gasoline to fill the tank. How many liters were in the tank before it was filled?

   _________ liters were in the tank.

2. Mr. Sumner can drive 5 kilometers on each liter of gasoline. How far could he drive on a full tank (85 liters) of gasoline?

   He could drive _________ kilometers on a full tank.

3. Miss Gray uses 17 liters of gasoline to drive to and from work each day. How many liters does she use in 6 days?

   She uses _________ liters in 6 days.

4. Evan bought 12 liters of paint. The paint was in 3 cans of the same size. How many liters of paint were in each can?

   _________ liters of paint were in each can.

5. Chelsea used 56 liters of water to fill 8 empty fishbowls. The same amount of water was in each bowl. How many liters were in each fishbowl?

   _________ liters of water were in each fishbowl.

6. A cafeteria serves 95 liters of milk each day. How much milk is served in 5 days?

   _________ liters of milk is served in 5 days.

7. Heidi uses 2 liters of gasoline to mow a lawn. She mowed the lawn 16 times this year. How much gasoline did she use to mow the lawn this year?

   Heidi used _________ liters this year.

1.	2.
3.	4.
5.	
6.	7.

NAME ____________________

CHAPTER 12 TEST

Find each length to the nearest centimeter.

1. ________ cm

2. ________ cm

3. ________ cm

Draw from A to B, from B to C, and from C to A.
Then find each length to the nearest centimeter.

C•

4. From A to B is _____ centimeters.

5. From B to C is _____ centimeters.

6. From C to A is _____ centimeters.

A• •B

How many liters would each container hold? Ring the best answer.

7.

1 liter 10 liters

8.

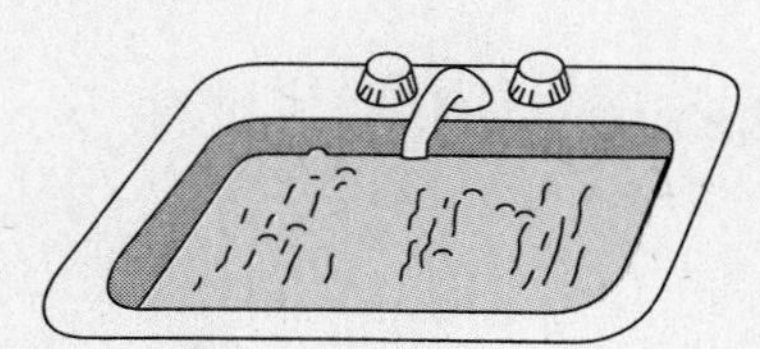

2 liters 20 liters

9.

2 liters 20 liters

Solve each problem.

10. A large carton holds 4 liters of milk. How many liters would 5 of these cartons hold?

    Five cartons would hold ________ liters.

11. Kelsey used an 8-liter sprinkling can to water some plants. She filled the can 4 times. How much water was used?

    ________ liters of water were used.

12. A car can go 9 kilometers on 1 liter of gasoline. How far could the car go on 50 liters?

    The car could go ________ kilometers.

10.	
11.	12.

12

PRE-TEST—Measurement

Find the length of each object to the nearest inch.

1. ________ inches

2. ________ inch

3. ________ inches

Complete the following.

	a	*b*
4.	1 quart = ________ pints	2 quarts = ________ pints
5.	8 pints = ________ quarts	6 pints = ________ quarts
6.	1 gallon = ________ quarts	3 gallons = ________ quarts
7.	8 quarts = ________ gallons	20 quarts = ________ gallons
8.	1 foot = ________ inches	3 feet = ________ inches
9.	1 yard = ________ feet	1 yard = ________ inches
10.	2 weeks = ________ days	1 hour = ________ minutes
11.	4 weeks = ________ days	6 hours = ________ minutes
12.	1 day = ________ hours	2 days = ________ hours

Solve.

13. Annette bought a board that is 6 feet long. What is the length of the board in inches?

It is ________ inches long.

NAME ______________________

Lesson 1 Inch

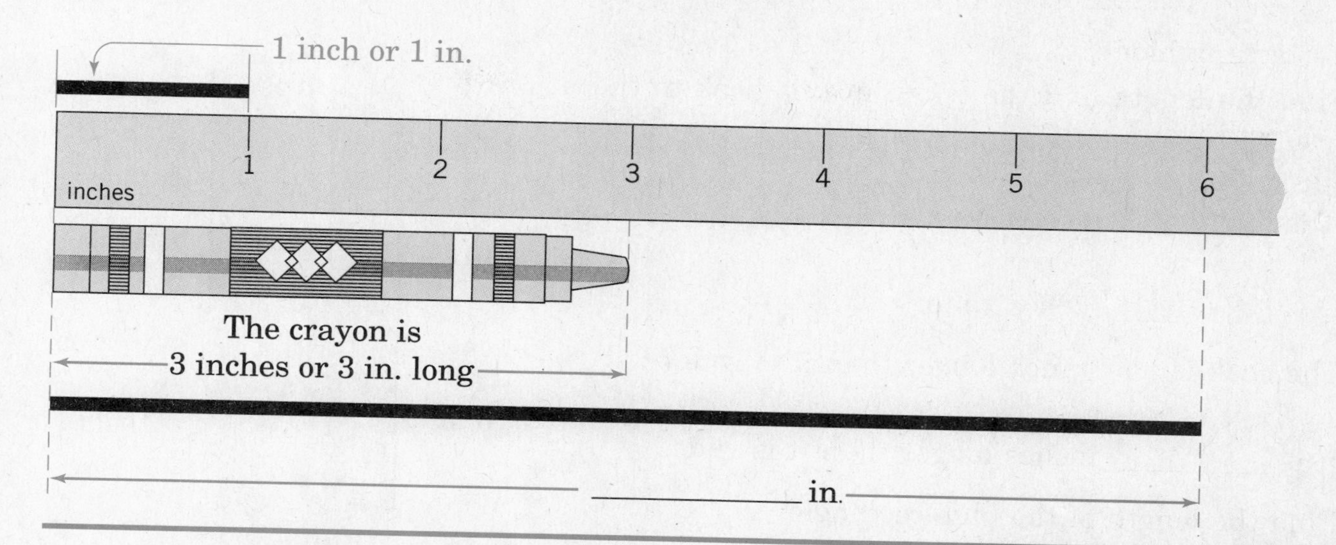

Find the length of each object to the nearest inch.

1. ______ in.

2. ______ in.

3. ______ in.

4. ______ in.

5. ______ in.

Complete the table.

	From	*Length*
6.	A to B	______ in.
7.	A to C	______ in.
8.	B to D	______ in.
9.	B to E	______ in.
10.	A to D	______ in.

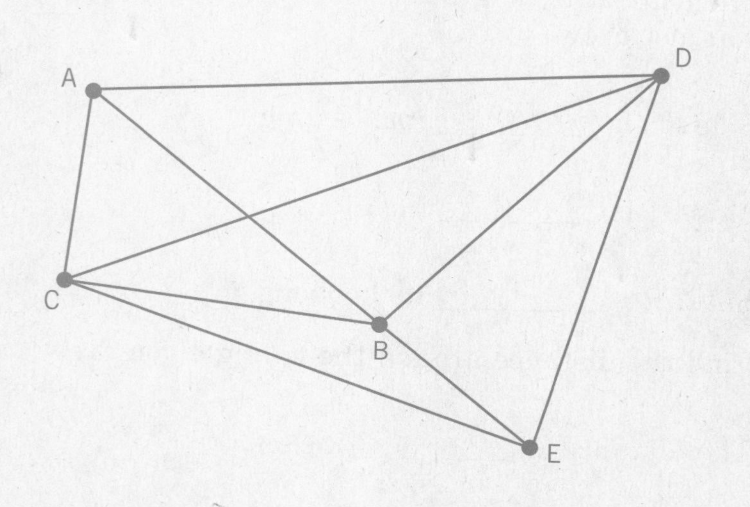

Problem Solving

Solve each problem.

1. Find the length and the width of this book to the nearest inch.

It is __________ inches long.

It is __________ inches wide.

2. The book is how much longer than it is wide?

It is __________ inches longer than it is wide.

3. Find the length of the blue rectangle.

It is __________ inches long.

4. Find the width of the blue rectangle.

It is __________ inch wide.

5. The rectangle is how much longer than it is wide?

It is __________ inches longer than it is wide.

6. Find the distance around the blue rectangle.

The distance is __________ inches.

7. Draw from R to S, from S to T, and from T to R. Then find the length of each side of the triangle you just drew.

R

Side RS is __________ inches long.

Side ST is __________ inch long.

Side TR is __________ inches long.

S • •T

8. Find the distance around the triangle you drew.

The distance is __________ inches.

NAME ______________________

Lesson 2 Measurement

1 foot (ft) = 12 inches (in.)

3 ft = ___?___ in.

Since 1 ft = 12 in., then

↓	↓
1	12
×3	×3
3	36
↓	↓

3 ft = 36 in.

3 feet (ft) = 1 yard (yd)

12 ft = ___?___ yd

Since 3 ft = 1 yd, then

$3\overline{)12}$ = 4

12 ft = 4 yd

Complete the following.

	a	*b*
1.	6 ft = ________ in.	8 ft = ________ in.
2.	3 yd = ________ ft	8 yd = ________ ft
3.	1 yd = ________ in.	2 yd = ________ in.
4.	21 ft = ________ yd	12 ft = ________ yd
5.	5 yd = ________ ft	14 yd = ________ ft
6.	3 ft = ________ in.	3 yd = ________ in.
7.	7 yd = ________ ft	9 ft = ________ in.
8.	18 ft = ________ yd	11 yd = ________ ft
9.	9 yd = ________ ft	5 ft = ________ in.
10.	7 ft = ________ in.	27 ft = ________ yd
11.	6 yd = ________ ft	15 ft = ________ yd
12.	9 ft = ________ yd	10 yd = ________ ft

Problem Solving

Solve each problem.

1. Teresa bought 2 yards of ribbon for a dress. How many feet of ribbon did she buy? Teresa bought __________ feet of ribbon.	**1.**
2. Myron bought a belt that was 2 feet long. How long was the belt in inches? The belt was __________ inches long.	**2.**
3. Mark has a rope that is 3 yards long. How long is the rope in feet? It is __________ feet long.	**3.**
4. In problem **3,** how long is the rope in inches? It is __________ inches long.	**4.**
5. Preston has a piece of wire 5 feet long. How long is the wire in inches? It is __________ inches long.	**5.**
6. The distance between 2 walls is 18 feet. What is this distance in yards? It is __________ yards.	**6.**
7. Pam's driveway is 15 feet wide. How wide is the driveway in yards? It is __________ yards wide.	**7.**
8. A fence post is 4 feet high. How high is the fence post in inches? It is __________ inches high.	**8.**

NAME ______

Lesson 3 Measurement

1 quart (qt) = 2 pints (pt)

5 qt = ? pt

Since 1 qt = 2 pt, then

$$\begin{array}{cc} 1 & 2 \\ \times 5 & \times 5 \\ \hline 5 & 10 \end{array}$$

5 qt = 10 pt

2 pint (pt) = 1 quart (qt)

10 pt = ? qt

Since 2 pt = 1 qt, then

10 pt = 5 qt

1 gallon (gal) = 4 qt or 4 qt = 1 gal

12 qt = ? gal

Since 4 qt = 1 gal, then

12 qt = 3 gal

Complete the following.

	a	*b*
1.	7 qt = ______ pt	3 qt = ______ pt
2.	8 pt = ______ qt	18 pt = ______ qt
3.	5 gal = ______ qt	2 gal = ______ qt
4.	24 qt = ______ gal	36 qt = ______ gal
5.	4 pt = ______ qt	7 gal = ______ qt
6.	8 qt = ______ pt	20 qt = ______ gal
7.	8 gal = ______ qt	12 pt = ______ qt
8.	28 qt = ______ gal	9 qt = ______ pt
9.	14 pt = ______ qt	9 gal = ______ qt

Problem Solving

Solve each problem.

1. Mrs. Collins bought 12 quarts of milk last week. How many pints of milk was this? It was __________ pints.	**1.**
2. In problem **1,** how many gallons of milk did Mrs. Collins buy? She bought __________ gallons.	**2.**
3. Mr. Murphy used 24 quarts of paint to paint his house. He bought paint in gallon cans. How many gallons of paint did he use? He used __________ gallons of paint.	**3.**
4. Mr. Johnson sold 18 pints of milk yesterday. How many quarts of milk was this? It was __________ quarts of milk.	**4.**
5. Dominic made 8 quarts of lemonade for a party. How many gallons of lemonade did he make? He made __________ gallons of lemonade.	**5.**
6. Patrick drank 10 pints of milk one week. How many quarts of milk did he drink? He drank __________ quarts of milk.	**6.**
7. Ms. Carlow used 4 gallons of paint. How many quarts of paint did she use? She used __________ quarts of paint.	**7.**
8. How many pint glasses could be filled from 8 quarts of juice? __________ pint glasses could be filled.	**8.**

NAME ______________________

Lesson 4 Time

3 weeks = ___?___ days

1 week = 7 days

↓	↓
1	7
×3	×3
3	21
↓	↓

3 weeks = ___21___ days

4 hours = ___?___ min

1 hour = 60 min

4 hours = ________ min

2 days = ___?___ hours

1 day = 24 hours

2 days = ________ hours

Complete the following.

	a	*b*
1.	2 weeks = ____________ days	8 weeks = ____________ days
2.	5 hours = ____________ min	7 hours = ____________ min
3.	6 days = ____________ hours	4 days = ____________ hours
4.	6 hours = ____________ min	9 weeks = ____________ days
5.	6 weeks = ____________ days	7 days = ____________ hours
6.	9 days = ____________ hours	3 hours = ____________ min
7.	9 hours = ____________ min	5 weeks = ____________ days
8.	8 days = ____________ hours	7 weeks = ____________ days

Problem Solving

Solve each problem.

1. Brad was at camp for 5 weeks. How many days was he at camp?

There are _________ days in 1 week.

He was at camp _________ weeks.

He was at camp _________ days.

1.

2. Tanya attends school 6 hours every school day. How many minutes does she attend every school day?

There are _________ minutes in 1 hour.

She attends school _________ hours.

She attends school _________ minutes.

2.

3. Holly was in the hospital for 4 days. How many hours was she in the hospital?

There are _________ hours in 1 day.

Holly was in the hospital _________ days.

She was in the hospital _________ hours.

3.

4. The Cooke family has lived in their new apartment for 6 weeks. How many days have they lived in their new apartment?

They have lived there _________ days.

4.

5. Mackenzie was away from home for 1 week. How many hours was she away from home?

Mackenzie was away from home _________ hours.

5.

NAME ____________________

Lesson 5 Problem Solving

Solve each problem.

1. A piece of wire is 2 feet long. How long is the wire in inches?

   The wire is ________ inches long.

2. If you use 14 inches of the wire in problem **1**, how many inches are left?

   There will be ________ inches left.

3. On a football field there are 100 yards between goal lines. How many feet is that?

   There are ________ feet between goal lines.

4. A container holds 8 quarts of liquid. How many pints does that container hold?

   That container holds ________ pints.

5. How many gallons of liquid does the container in problem **4** hold?

   That container holds ________ gallons.

6. A telethon lasted 2 days. How many hours did the telethon last?

   The telethon lasted ________ hours.

7. A television mini-series lasted 6 hours. How many minutes did the mini-series last?

   The mini-series lasted ________ minutes.

8. The Mohrs spent 3 weeks on their vacation trip. How many days was that?

   The vacation trip took ________ days.

1.	**2.**
3.	**4.**
5.	**6.**
7.	**8.**

NAME ____________________

CHAPTER 13 TEST

Find each length to the nearest inch.

1. ________ in.

2. ________ in.

3. ________ in.

4. ________ in.

Complete the following.

	a	*b*
5.	4 ft = ________ in.	4 pt = ________ qt
6.	6 qt = ________ pt	21 ft = ________ yd
7.	8 gal = ________ qt	9 ft = ________ in.
8.	9 yd = ________ ft	36 qt = ________ gal
9.	6 weeks = ________ days	8 hours = ________ min
10.	3 days = ________ hours	8 weeks = ________ days

Solve each problem.

11. Kaylee has a rope 8 feet long. How long is the rope in inches?

It is ________ inches long.

11.

12. Craig had a gallon of gasoline. He used 1 quart for the lawn mower. How many quarts did he have left?

He had ________ quarts left.

12.

13. Tonya is 4 feet 11 inches tall. What is her height in inches?

Her height is ________ inches.

13.

NAME ____________________

TEST Chapters 1-6

Add or subtract.

	a	b	c	d	e
1.	6 +5	9 +9	61 +5	35 +43	9 +42
2.	84 +7	64 +18	70 +70	18 +93	85 +89
3.	70 80 +30	478 +596	256 175 +310	$6.98 +7.23	$11.62 3.65 +1.98
4.	16 −7	15 −6	38 −4	87 −15	56 −8
5.	43 −7	80 −17	150 −70	136 −69	181 −93
6.	876 −97	625 −208	8724 −893	$9.85 −6.27	$18.20 −6.75

Test CH. 1-6

For each clockface, write the numerals that name the time.

7.

a

_____ : _____

b

_____ : _____

c

_____ : _____

Solve each problem.

8. There are 9 bolts in one package. There are 6 bolts in another package. How many bolts are in both packages?

There are __________ bolts in both packages.

9. There are 27 letters to be typed. Only 6 letters have been typed. How many letters still need to be typed?

__________ letters still need to be typed.

10. This morning the temperature was 58°F. Now it is 18° warmer. What is the temperature now?

Now the temperature is __________°F.

11. Dirk read 84 pages in the morning. He read 69 pages in the afternoon. How many pages did he read that day?

He read __________ pages that day.

12. There were 103 plants in the garden. Only 8 were bean plants. How many were not bean plants?

__________ were not bean plants.

13. You bought items at a store that cost $1.45, $2.98, and $9.98. How much did these items cost altogether?

These items cost $__________ altogether.

14. Jennifer wants to buy a purse that costs $18.29. She has $9.55. How much more does she need to buy the purse?

She needs $__________ more.

8.	**9.**
10.	**11.**
12.	**13.**
14.	

NAME ______________________________

FINAL TEST Chapters 1–13

Add or subtract.

	a	*b*	*c*	*d*	*e*
1.	$\begin{array}{r} 7 \\ +8 \\ \hline \end{array}$	$\begin{array}{r} 65 \\ +3 \\ \hline \end{array}$	$\begin{array}{r} 82 \\ +16 \\ \hline \end{array}$	$\begin{array}{r} 26 \\ +9 \\ \hline \end{array}$	$\begin{array}{r} 73 \\ +19 \\ \hline \end{array}$
2.	$\begin{array}{r} 20 \\ +90 \\ \hline \end{array}$	$\begin{array}{r} 69 \\ +43 \\ \hline \end{array}$	$\begin{array}{r} 165 \\ +927 \\ \hline \end{array}$	$\begin{array}{r} 36 \\ 72 \\ +35 \\ \hline \end{array}$	$\begin{array}{r} \$21.43 \\ +62.97 \\ \hline \end{array}$
3.	$\begin{array}{r} 12 \\ -4 \\ \hline \end{array}$	$\begin{array}{r} 56 \\ -5 \\ \hline \end{array}$	$\begin{array}{r} 93 \\ -83 \\ \hline \end{array}$	$\begin{array}{r} 62 \\ -7 \\ \hline \end{array}$	$\begin{array}{r} 85 \\ -47 \\ \hline \end{array}$
4.	$\begin{array}{r} 180 \\ -90 \\ \hline \end{array}$	$\begin{array}{r} 125 \\ -76 \\ \hline \end{array}$	$\begin{array}{r} 780 \\ -539 \\ \hline \end{array}$	$\begin{array}{r} 3751 \\ -865 \\ \hline \end{array}$	$\begin{array}{r} \$25.00 \\ -7.25 \\ \hline \end{array}$

Multiply.

5.	$\begin{array}{r} 5 \\ \times 3 \\ \hline \end{array}$	$\begin{array}{r} 4 \\ \times 4 \\ \hline \end{array}$	$\begin{array}{r} 6 \\ \times 5 \\ \hline \end{array}$	$\begin{array}{r} 8 \\ \times 3 \\ \hline \end{array}$	$\begin{array}{r} 7 \\ \times 2 \\ \hline \end{array}$
6.	$\begin{array}{r} 3 \\ \times 9 \\ \hline \end{array}$	$\begin{array}{r} 8 \\ \times 7 \\ \hline \end{array}$	$\begin{array}{r} 6 \\ \times 8 \\ \hline \end{array}$	$\begin{array}{r} 9 \\ \times 7 \\ \hline \end{array}$	$\begin{array}{r} 6 \\ \times 6 \\ \hline \end{array}$

Multiply.

	a	*b*	*c*	*d*
7.	$\begin{array}{r} 70 \\ \times 5 \\ \hline \end{array}$	$\begin{array}{r} 23 \\ \times 3 \\ \hline \end{array}$	$\begin{array}{r} 18 \\ \times 7 \\ \hline \end{array}$	$\begin{array}{r} 43 \\ \times 7 \\ \hline \end{array}$

Divide.

8.	$2\overline{)12}$	$4\overline{)20}$	$3\overline{)27}$	$1\overline{)8}$
9.	$5\overline{)40}$	$4\overline{)24}$	$7\overline{)0}$	$8\overline{)72}$
10.	$9\overline{)36}$	$7\overline{)49}$	$6\overline{)48}$	$9\overline{)54}$

Answer each question. Use the calendar to help you.

June

S	M	T	W	T	F	S
				1	2	3
4	5	6	7	8	9	10
11	12	13	14	15	16	17
18	19	20	21	22	23	24
25	26	27	28	29	30	

11. How many days are in June? __________

12. On what day is June 17? __________

Go to next page.

NAME ______________________

FINAL TEST (continued)

13. Write the numerals that name the time.

_______ : _______

Find the length to the nearest centimeter.

14. _______ centimeters

Find the length to the nearest inch.

15. _______ inches

Solve each problem.

16. Juan has 36 inches of wire. How many feet of wire does he have?

Juan has _______ feet of wire.

17. Caryl used 3 liters of gasoline to mow a lawn. She mowed the lawn 12 times this year. How much gasoline did she use to mow the yard this year?

She used _______ liters this year.

16.

17.

Solve each problem.

18. Amanda worked in the automobile factory for 45 days. She worked 5 days each week. How many weeks did she work?

Amanda worked _________ weeks.

19. One store sold 421 radios. Another store sold 294 radios. A third store sold 730 radios. How many radios did all three stores sell?

All three stores sold _________ radios.

20. Ms. O'Connor received 1,439 votes. Mr. Ortega received 810 votes. How many more votes did Ms. O'Connor receive than Mr. Ortega?

Ms. O'Connor received _________ more votes.

21. Branden bought 24 liters of paint. The paint was in 6 cans. All the cans were the same size. How many liters of paint were in each can?

There were _________ liters of paint in each can.

22. A farmer has 6 stacks of bales of hay. Each stack has 36 bales. How many bales of hay does the farmer have?

The farmer has _________ bales of hay.

23. Larysa bought items that cost $3.80, $2.29, and $1.75. The sales tax was $0.47. How much did she pay for those items, including tax?

She paid $_________.

24. Don had $27.89. He spent $11.92. How much did he have left?

Don had $_________ left.

18.	**19.**
20.	**21.**
22.	**23.**
24.	

NAME ______________________

CHAPTER 1 CUMULATIVE REVIEW

Add.

	a	*b*	*c*	*d*	*e*	*f*
1.	3 +8	6 +5	9 +1	5 +7	6 +4	2 +9
2.	6 +7	3 +9	9 +8	7 +4	6 +8	9 +9
3.	8 +5	1 +9	7 +8	6 +6	3 +7	9 +6
4.	8 +8	4 +8	5 +5	9 +7	7 +7	2 +8

Subtract.

	a	*b*	*c*	*d*	*e*	*f*
5.	13 −6	11 −9	14 −7	10 −3	13 −4	15 −7
6.	10 −6	12 −3	14 −6	11 −7	15 −8	10 −2
7.	14 −8	11 −6	10 −5	16 −7	12 −8	17 −9
8.	16 −8	17 −9	15 −6	13 −8	17 −8	18 −9

NAME ____________________

CHAPTERS 2 and 3 CUMULATIVE REVIEW

Add or subtract. Check each answer.

	a	*b*	*c*	*d*	*e*	*f*
1.	5 + 3	36 + 2	45 + 3	7 + 31	56 + 3	8 + 40
2.	42 + 13	56 + 23	33 + 64	51 + 23	60 + 28	35 + 44
3.	45 + 8	56 + 7	9 + 64	16 + 6	7 + 55	48 + 7
4.	29 + 46	27 + 39	36 + 18	45 + 26	18 + 38	39 + 35
5.	9 − 6	46 − 3	38 − 7	55 − 2	59 − 8	67 − 4
6.	65 − 32	78 − 26	85 − 71	96 − 72	74 − 33	84 − 30
7.	35 − 7	46 − 9	51 − 8	24 − 7	63 − 6	72 − 9
8.	84 − 39	75 − 26	63 − 18	92 − 47	51 − 35	76 − 28

NAME ____________________

CHAPTERS 4 and 5 CUMULATIVE REVIEW

Add or subtract.

	a	b	c	d	e
1.	50 +70	63 +26	80 +72	98 +16	56 +46
2.	6 7 +8	15 4 +30	26 75 +48	64 98 +56	50 75 +24
3.	675 432 +109	556 374 +897	57 25 43 +86	516 709 823 +674	780 506 392 +667
4.	160 −70	140 −90	164 −81	132 −94	146 −78
5.	789 −362	370 −164	1708 −485	7534 −278	3650 −957

Solve each problem.

6. Last month, Jason received four paychecks in the following amounts: $675, $593, $607, and $753. What was the total amount Jason earned last month?

Last month Jason earned __________.

6.

7. Jordan has $257 in her savings account. Patty has $1,025 in her account. How much more does Patty have in her account than Jordan?

Patty has __________ more.

7.

NAME ____________________

CHAPTER 6 CUMULATIVE REVIEW

For each clockface, write the numerals that name the time.

	a	*b*	*c*
1.	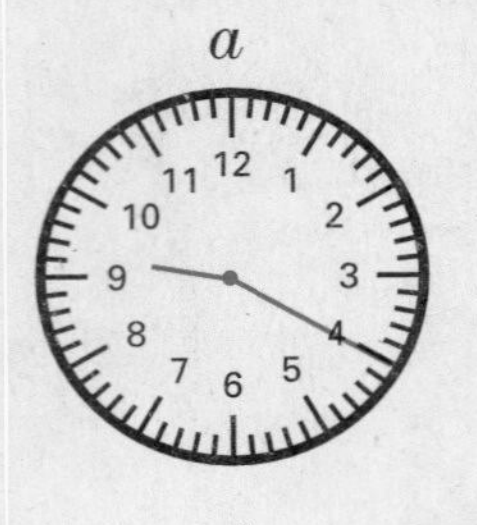		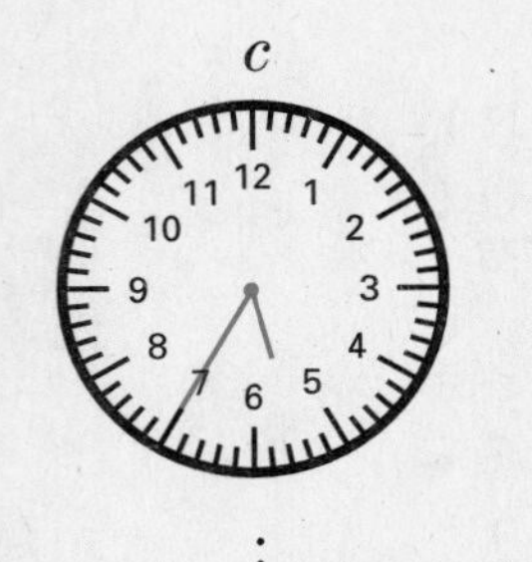
	______ : ______	______ : ______	______ : ______

Complete the following.

	a	*b*	*c*
2.	XIV = ________	XXV = ________	XXXIII = ________
3.	16 = ________	24 = ________	39 = ________

Add or subtract.

	a	*b*	*c*	*d*	*e*
4.	\$1 8.3 5 +2 1.4 2	\$9.6 7 +0.3 2	3 6¢ +4 2¢	1 8¢ 2 5¢ +3 0¢	\$1 5.7 3 0.4 9 +2.2 0
5.	\$5.4 1 −1.2 8	\$1 6.0 2 −7.8 1	7 2¢ −2 7¢	\$1 3.4 0 −2.0 8	\$8.1 2 −2.0 7

Solve.

6. Joseph has saved \$57.25 to buy a bicycle. The bicycle costs \$95.75. How much more money does Joseph need to buy the bicycle?

Joseph needs ________________ more.

6.

NAME ___________________

CHAPTERS 7 and 8 CUMULATIVE REVIEW

Multiply.

	a	*b*	*c*	*d*	*e*
1.	1×5	4×2	3×6	7×1	2×8
2.	5×3	6×4	9×2	3×7	4×5
3.	8×3	6×7	3×9	4×8	0×9
4.	7×7	8×6	9×1	5×6	8×8
5.	6×6	0×5	9×8	6×9	8×7

Solve.

6. Alexandria can ride her bicycle 4 miles in one hour. How far can she ride in 3 hours?

Alexandria can ride __________ miles.

6.

NAME ____________________

CHAPTER 9 CUMULATIVE REVIEW

Multiply.

	a	b	c	d	e
1.	$\begin{array}{r} 40 \\ \times 3 \\ \hline \end{array}$	$\begin{array}{r} 23 \\ \times 5 \\ \hline \end{array}$	$\begin{array}{r} 35 \\ \times 7 \\ \hline \end{array}$	$\begin{array}{r} 56 \\ \times 4 \\ \hline \end{array}$	$\begin{array}{r} 68 \\ \times 9 \\ \hline \end{array}$
2.	$\begin{array}{r} 27 \\ \times 8 \\ \hline \end{array}$	$\begin{array}{r} 39 \\ \times 2 \\ \hline \end{array}$	$\begin{array}{r} 46 \\ \times 6 \\ \hline \end{array}$	$\begin{array}{r} 73 \\ \times 8 \\ \hline \end{array}$	$\begin{array}{r} 82 \\ \times 4 \\ \hline \end{array}$
3.	$\begin{array}{r} 90 \\ \times 7 \\ \hline \end{array}$	$\begin{array}{r} 58 \\ \times 6 \\ \hline \end{array}$	$\begin{array}{r} 39 \\ \times 3 \\ \hline \end{array}$	$\begin{array}{r} 87 \\ \times 5 \\ \hline \end{array}$	$\begin{array}{r} 43 \\ \times 9 \\ \hline \end{array}$
4.	$\begin{array}{r} 51 \\ \times 6 \\ \hline \end{array}$	$\begin{array}{r} 60 \\ \times 7 \\ \hline \end{array}$	$\begin{array}{r} 92 \\ \times 8 \\ \hline \end{array}$	$\begin{array}{r} 76 \\ \times 4 \\ \hline \end{array}$	$\begin{array}{r} 81 \\ \times 3 \\ \hline \end{array}$
5.	$\begin{array}{r} 45 \\ \times 9 \\ \hline \end{array}$	$\begin{array}{r} 39 \\ \times 7 \\ \hline \end{array}$	$\begin{array}{r} 16 \\ \times 2 \\ \hline \end{array}$	$\begin{array}{r} 70 \\ \times 8 \\ \hline \end{array}$	$\begin{array}{r} 44 \\ \times 4 \\ \hline \end{array}$

Solve.

6. Mr. Garcia ordered 7 buses for a field trip. There were 36 students on each bus. How many students were there in all on the field trip?

There were ____________ students in all.

6.

NAME ______________________

CHAPTER 10 CUMULATIVE REVIEW

Divide.

	a	*b*	*c*	*d*	*e*
1.	$1\overline{)5}$	$3\overline{)6}$	$2\overline{)8}$	$4\overline{)28}$	$5\overline{)30}$
2.	$2\overline{)18}$	$5\overline{)40}$	$4\overline{)12}$	$3\overline{)3}$	$1\overline{)4}$
3.	$3\overline{)9}$	$1\overline{)3}$	$2\overline{)16}$	$5\overline{)45}$	$4\overline{)32}$
4.	$5\overline{)35}$	$4\overline{)20}$	$1\overline{)2}$	$3\overline{)27}$	$2\overline{)14}$
5.	$4\overline{)36}$	$1\overline{)1}$	$2\overline{)12}$	$5\overline{)25}$	$3\overline{)21}$
6.	$1\overline{)6}$	$3\overline{)24}$	$5\overline{)20}$	$2\overline{)10}$	$4\overline{)24}$

Solve.

7. There are 32 students in a class. Desks are arranged in groups of 4. How many groups of desks are there in the class?

There are __________ desks in all.

Desks are arranged in groups of __________.

There are __________ groups of desks in the class.

7.

NAME ______________________

CHAPTER 11 CUMULATIVE REVIEW

Divide.

	a	*b*	*c*	*d*	*e*
1.	$7\overline{)21}$	$6\overline{)30}$	$9\overline{)36}$	$8\overline{)56}$	$6\overline{)36}$
2.	$8\overline{)40}$	$7\overline{)28}$	$9\overline{)81}$	$6\overline{)54}$	$7\overline{)49}$
3.	$6\overline{)48}$	$9\overline{)72}$	$8\overline{)48}$	$7\overline{)56}$	$8\overline{)72}$
4.	$9\overline{)27}$	$7\overline{)63}$	$6\overline{)42}$	$8\overline{)16}$	$9\overline{)63}$
5.	$7\overline{)35}$	$6\overline{)24}$	$8\overline{)0}$	$9\overline{)45}$	$6\overline{)12}$
6.	$8\overline{)24}$	$9\overline{)54}$	$6\overline{)18}$	$7\overline{)42}$	$1\overline{)8}$

Solve.

7. Tina collects baseball cards. She has 54 cards. She wants to put them in containers that hold 6 cards each. How many containers does she need?

She has __________ cards in all.

Each container holds __________ cards.

She needs __________ containers.

7.

NAME ______________________

CHAPTER 12 CUMULATIVE REVIEW

Find each length to the nearest centimeter.

1. ________ cm

2. ________ cm

3. ________ cm

Complete the table to the nearest centimeter.

	From	*Length*
4.	A to B	________ cm
5.	B to C	________ cm
6.	E to D	________ cm
7.	D to B	________ cm
8.	A to C	________ cm

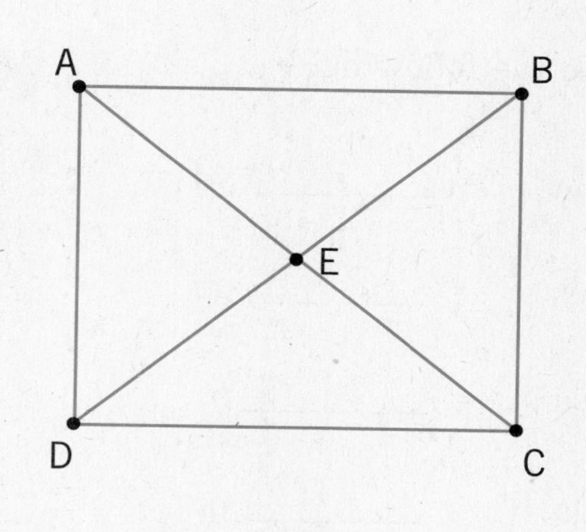

How many liters would each container hold?
Circle the best answer.

9. gallon jar	5 liters	20 liters	75 liters
10. kitchen sink	2 liters	5 liters	20 liters
11. bucket	3 liters	15 liters	40 liters
12. quart bottle	1 liter	10 liters	20 liters

NAME ____________

CHAPTER 13 CUMULATIVE REVIEW

Find each length to the nearest inch.

1. ________ in.

2. ________ in.

3. ________ in.

4. ________ in.

Complete the following.

	a		*b*	
5.	9 ft = ________ yd		5 gal = ________ qt	
6.	10 pt = ________ qt		6 ft = ________ in.	
7.	5 yd = ________ ft		2 yd = ________ in.	
8.	3 gal = ________ qt		12 qt = ________ gal	
9.	2 hours = ________ min		8 weeks = ________ days	
10.	5 days = ________ hours		5 hours = ________ min	

Solve.

11. Juan is 5 feet tall. How tall is he in inches?

Juan is ________ inches tall.

11.

12. Maria is 4 feet 10 inches tall. How tall is she in inches?

Maria is ________ inches tall.

12.